Dedication

This book is dedicated to my dear wife, the mother of our delightful six-year-old. Her joyous heart, loyal spirit and unconditional love made finding the solutions presented in these pages an imperative for me.

Acknowledgements

I owe a debt of gratitude to Dr. Jason Abber, my surgeon at the Southern California Permanente Group of the Kaiser Foundation Hospital. His skilled hands took me through an amazingly successful prostate operation, and in his office I began my search for ways to overcome erectile dysfunction.

Special acknowledgements to Dr. James A. Bolton whose foreword to this book clearly shows that he has a broad and wise understanding of the delicate role of erectile functioning in the bonding of husband and wife. His generous participation, in my project to present solutions for overcoming erectile dysfunction, was key in making this book possible. His sage advice and continuing support are invaluable.

I'm also grateful to all those who have discovered and created the many options now available (which I've covered in this book) to help men restore, maintain and enhance healthy erectile functioning.

And finally, I appreciate my publisher and editor for working closely with me to assure that this book conveyed my own personal struggle and victory.

About The Author

Brian Drew, former marathon runner, father, educator and now head of an educational institution, has personally overcome erectile dysfunction (ED). His prostate had been removed to save his life—leaving him impotent and therefore, unable to function sexually. He knows, firsthand, how devastating that can be for a man and his wife.

After his successful operation in 1995, he immersed himself in the subject, and then wrote this book. It is his desire to share what he learned during the process of regaining his potency, enabling him to completely restore his "love affair" with his wife.

As an educational leader, he's accustomed to doing extensive research, and cares about the welfare of others. This book reflects his careful, compassionate, well-thought-out approach to this most common, untreated and challenging of all male medical conditions. He sincerely hopes this book will be of great benefit to the men who read it, as well as to their wives, and to all whose lives are affected, directly or indirectly.

The author also hopes that the various solutions offered will help foster healthier, happier marriages. Brian, himself, enjoys such a marriage and lives happily with his wife, a registered nurse, and two sons, in California.

VIAGRA AND THE QUEST FOR POTENCY

Answers and Solutions to Restore and Enhance Your Quality of Life

BRIAN DREW with a Foreword by DR. JAMES A. BOLTON, PhD, MFCC

A Health Publishers Book

HEALTH PUBLISHERS
Distributed by
BookWorld Services, Inc.
Sarasota, Fl
1-800-444-2524
and INGRAM

Health Publishers is an imprint and the heart symbol with the Rx is a trademark of Markowski Int'l Publishers.

1 2 3 4 5 6 7 8 9 10

Library of Congress Cataloging-in-Publication Data
Drew, Brian
Viagra and the quest for potency: answers and solutions to restore and enhance your quality of life/Brian Drew· with a foreword
by James A. Bolton
1st ed
p. cm.
Includes bibliographical references and index

1.Drew, Brian – Health. 2. Impotence – Popular works. 3.Sildenafil – Therapeutic use – Popular works. 4. Masculinity (Psychology). 5. Sex instruction for men.
I. Title
RC889.D74 1998
616.6'92
QBI98-819
Preassigned LCCN: 98-66739
ISBN: 0-938716-55-7

A Word of Caution

This book is for information and education purposes only. Impotence, ie, erectile dysfunction (ED), is a serious medical condition that you need to discuss with your doctor. Do not attempt to diagnose or treat yourself, and don't borrow any treatments or medications from anyone else. Always consult with your physician before embarking on any health program or change of habits. The publisher and author cannot be held responsible for your results or lack of results. No liability is assumed with respect to use of the information herein. All names cited in case histories are fictitious to protect their confidentiality. Also, some of the circumstantial information may have been altered somewhat. Any similarity found in these stories that may fit you or someone you know is purely coincidental.

Manufactured in the United States of America

Contents

"The one law that does not change
is that everything changes,
and the hardship I was bearing today
was only a breath away from
the pleasures I would have tomorrow,
and those pleasures would be
all the richer because of the memories
of this I was enduring."

Louis L'Amour

Foreword

By Dr. James A. Bolton, Ph.D., Marriage & Family Counseling

"It's unlikely that we can overestimate the importance of a satisfactory sex life to having an overall sense of well-being."

When Brian Drew began preparing this book, neither he nor I, or his wife, realized how much progress would be made in such a short time, in curing erectile dysfunction (ED). In the spring of 1998, we saw headlines around the world emblazoned with news about the new "Impotence Pill."

This pill is called Viagra which is Pfizer's trade name for sildenafil citrate—the active ingredient. By all accounts, and by the experience of hundreds of thousands of men, this is certainly very nearly that aphrodisiac-like substance men and women have sought and speculated about throughout recorded history. It may come as a surprise to many that several other reliable methods of overcoming ED (historically called impotence) had already been developed and made available before the introduction of Viagra!

In my practice as a marriage and family counselor, I have treated hundreds of distraught patients, in which the cause of their unhappiness was ultimately traced to lack of sexual fulfillment in the intimacy of their marriage. It's unlikely that we can overestimate the importance of a satisfactory sex life to having an overall sense of well-being. Furthermore, wives often blame themselves for their husband's failure to attain an erection during intimacy. A situation like this can cast a shadow over the entire relationship—often driving both parties into grievous or angry silence and even divorce.

For the fortunate, sex is an indispensable and integral part of their lives. But, in many cases, the success or failure of lovemaking between a man and a woman is directly traceable to the man's ability, or lack of it, to achieve an erection. Unfortunately, the incidence of erection failure is of epidemic proportion. This is not to say, however, that every man with an erection problem has it all the time. Many, if not most, have only passing episodes of erection failure but too many others have continual, chronic failure. Taken all together however, it amounts to an epidemic.

It must be added that some erection problems are not as bad as others. There are many men who simply feel they do not get as hard as they think they should or as they once did. In any case, it is now possible, in nearly all cases, to overcome erection problems. Luckily, we live in a time when great strides have been made in understanding the mechanics of male sexual arousal and erectile functioning, as well as in finding ways to overcome the inability to have a satisfactory erection.

This book is for men who would like to restore, enhance or maintain their sexual functioning. It's also for those who think they don't get as "hard" as they should, and especially for those men with chronic erection failure.

I also suggest that women read this book too, so they can better understand and support their man in achieving and maintaining excellent sexual health.

Information is now available to help you improve the quality of your erections. This book fills an important need by presenting what you need to know in a clear, concise, organized form. The therapies discussed here are virtually identical to those I use in my practice, and many are being prescribed by doctors. This includes the "miracle" drug, Viagra, which received FDA (the U.S. Food and Drug Administration) approval in late March, 1998. This marvelous pharmaceutical is proclaimed to be an almost instantaneous cure for impotence.

Remarkably, other similar treatments (also discussed in this book) are now undergoing clinical trials, as of this writing. Many are already being prescribed by physicians throughout the world. I am confident that they, too, will soon be available with

FDA approval to market them as cures for impotence.

In my own clinical practice, I work with urologists. When necessary, they perform medical examinations and prescribe Viagra and other pharmaceuticals, as appropriate, for my patients. Like many other doctors of this specialty, they also prescribe other treatments, including some discussed in this book. These include prescription yohimbe, vacuum therapy, self-injected pharmaceuticals, and such. Consequently, I feel quite confident in giving this much needed book a hearty welcome and an unqualified endorsement.

It is my opinion that, if you are a male in reasonably good health, with at least a modest ability at self-discipline, the information presented in this book can help you—possibly tremendously. I would even go so far as to say that if this information cannot help you, then it is most likely that drastic medical procedures will be required. But those situations would be rare.

The author has firsthand knowledge about the most advanced practices included in many aspects of the medical treatment of impotence. He is also well-grounded in alternative health practices and medical treatments. Some of these "alternatives" have proven to be quite effective even though they are, generally speaking, not prescribed by the medical establishment. In fact, many medical doctors may not even know of them!

This situation however, is changing, as more and more research by the medical establishment uncovers the effectiveness of "health food store" natural remedies. One of the best examples indicating that a shift in attitude is now occurring is the widespread attention and respect being paid to the well-known physician and bestselling author of *Spontaneous Healing*, Dr. Andrew Weil. He is a leading advocate of the acceptance of natural remedies that have been established through actual use, as effective.

This book grew out of the author's desire to share information about solutions to overcoming ED with other men who are experiencing erection problems. The author was well prepared for this task. He, himself, had wrestled with the problem of ED as an unfortunate outcome of a radical prostatectomy. Assisted by his urologist, he explored not only the solutions offered by the traditional medical profession but also those of alternative medicine. As a result of Brian's research and his own firsthand knowledge and experience, millions of men will benefit. However, the best result is that the findings and experiences of the author are preserved and presented in this book. It is hoped this book will benefit many others, not only the men who will use this information, but also the women who love them!

– Dr. James A. Bolton, Ph.D., MFCC

Introduction

"The great news is that there's help for nearly every degree of ED (erectile dysfunction) there is—you can now do something about it."

Do You Want To Be More Potent?

The purpose of this book is to inform men, and the women who love them, so they can enjoy the fullest potential of their lovemaking. If you are among the majority of men who wish to improve their lovemaking performance, this book is for you. Also, if you are among the countless number of men who experience erection problems, this book is certainly for you. Or, if you just want to maintain the quality of your sexual functioning, this book can help you.

Perhaps you're a frustrated wife whose husband continually, or even fairly regularly lets you down because he fails to "perform;" you can help him overcome this situation by reading this book.

In the June 11, 1997 edition of the *Los Angeles Times,* it was reported that as many as 30 million men, in the U.S. alone, experience occasional impotence. Moreover, an astounding 50% of men aged 40 to 70 experience problems maintaining an erection. The worst news, according to the report, is that 18 million men, just in the U.S., can be classified as impotent.[1]

The term "impotence" was established through use by the medical profession to refer to "erectile dysfunction"(ED). It is an unfortunate choice of terms since its root meaning is "without power." In fact, the problem being addressed has little to do with power. Strong, powerful men can have ED problems, just as much as anyone else! And, as you will learn, even men with ED problems can have powerful sex. Therefore, the term "erectile dysfunction" is a more accurate term than the term "impotence."

Who Gets ED?

Many people believe that ED is a natural consequence of aging. This just isn't true! It is true that more erection problems are found among older men. However, this is most often a side effect of other health problems and did not occur as a natural consequence of getting older. So, while the probability of ED increases with age, fortunately enough, it's not an inevitable consequence of aging.

Another common belief is that most men occasionally experience erection problems. This is not scientifically proven but very likely to be true. Without even considering the adverse effects of certain serious medical problems, like diabetes and a rather long list of medications on erectile functioning, a man's ability to make love can be diminished by a number of factors. Some of them are part of the common experiences of everyday life. They include what he eats and drinks, stress, fatigue, illness, worry and fear. So, occasionally, these seemingly innocent experiences or factors of normal living are the cause of a man's inability to achieve an erection. But, these kinds of ED problems are not nearly as bothersome or devastating as chronic, persistent erection failure or weak or partial erections.

"Potency-On-Demand"—Can Everyone Have A Better Sex Life?

The great news is that there's help for nearly every degree of ED there is—you can now do something about it. You owe it to yourself to explore the possibilities. A full and exhilarating sex life can add zest to each day of your life and whether you have it or not is basically up to you. By investing in yourself through purchasing this book, you have set out on the road to achieving your full potential for sexual expression as a male. And I am confident you can achieve this goal by applying the information I'll be sharing with you here.

There are methods available today for overcoming erection

difficulties that would have been unbelievable just a decade ago. Men are now able to witness, right before their eyes, the almost-instantaneous restoration of their erectile functioning. As one amazed recipient of these methods said, "It's like getting an erection-on-demand!"

In this book, you will learn about the latest scientific developments in treating ED. The information is presented with sufficient explanation and detail to enable you to decide whether a particular treatment is of interest to you. Of the various approaches, only those involving surgeries are not covered in any detail in this book, including penile implants and surgical repair of leaking penile vascular tissue. You can discuss these methods with your physician. In addition, reports from case histories are presented illustrating applications of these (nonsurgical) approaches. It is now possible for a man to design his own program for achieving better erections, by selecting what method or combination of methods would work best and appeal most to him. *Before you do anything, though, be sure to consult with your doctor.*

"Impotence" is a medical condition you can do something about, and this book brings to you the information you need. It will be very helpful to men whose goal is simply to improve the firmness of their erections as well as to men who wish to overcome persistent erection failure. Achieving an erection need no longer be left entirely to nervous chance.

I Couldn't Believe It Was Happening To Me!

My interest in this subject grew out of the prospect that my own potency was threatened by prostate cancer. When my urologist told me that he needed to do a biopsy on my prostate gland because a routine blood test (a prostate specific antigen–PSA test) had revealed that I might have prostate cancer, I was in a state of disbelief. It took two years of examining tiny tissue samples of my prostate gland, before this very able urologist was able to confirm the correctness of the dreaded diagnosis he suspected. In the meantime—all the while, in fact, I fiercely resisted the possibility that such a diagnosis might be true.

To me, my prostate was working just fine. It never gave me any trouble, it wasn't enlarged as is often the case with men in my age group, and it seemed to be doing what it was supposed to do. Indeed, I was the father of a toddler boy and I was already sixty. Moreover, I thought of myself as healthy. I had even completed several marathons in the five years immediately preceding the diagnosis. I exercised regularly and basically practiced good health habits. How could this be happening to me?

I Began Asking Questions And Searching For Answers

After I finally accepted the diagnosis, it soon became clear to me that the only rational course of action for a man in my predicament was to let the surgeon perform the recommended radical excision. Realizing that prostate surgery might leave my erectile functioning seriously impaired, I began reading, searching and asking questions. The result is this book.

I Found Some Helpful Information

I was already familiar with much of the traditional lore and wisdom available from alternative medicine sources. I was, in fact, a regular customer at a local health food store.

More importantly, though, I was strongly influenced by the message of Dr. Andrew Weil about the human body's innate healing abilities and the limitations of medicine. His willingness to accept natural healing sources seemed very wise to me, since I personally experienced benefits from using natural healing herbs and other substances. His newsletter and books are a source of deep insights, inspiration and sound advice.

As I faced that most radical threat to my erectile functioning, prostate surgery, I found his book, *Spontaneous Healing*. It served as a source of inspiration to me, and my decisions about my impending surgery were influenced by the advice I obtained in its pages. My health decisions continue to be influenced by it. Since the publication of *Spontaneous Healing*, Americans and people throughout the world are paying more attention to the messages of alternative medicine.

In August 1997 a book was published that men concerned about erectile functioning might want to examine. Larry Clapp's *Prostate Health in 90 Days Without Drugs or Surgery* is solidly in the camp of alternative medicine and deals with a subject intrinsically involved in erectile functioning.

Another author and physician gaining wide recognition for his advocacy of alternative medicine and health advice is Dr. Isadore Rosenfeld. Thousands read his articles, "A Doctor's Guide to Herbs," which appear from time to time in *Parade Magazine,* distributed mainly with your Sunday newspaper. Most importantly, this medical doctor is the author of *Dr. Rosenfeld's Guide to Alternative Medicine: What Works, What Doesn't and What's Right for You.*

I Discovered Various Remedies Are Available

There are natural healing therapies for overcoming ED and if they work for you, that's great, but always consult with your doctor; don't take any chances. If they do not, you live at a most fortunate time in the history of medicine and medicinal technology. Through my urologist I learned of amazing therapies for overcoming ED. By using them, I overcame ED, just as any reader of this book should be able to do if he needs or desires to—unless, of course, he has a serious medical condition that cannot be compensated for.

I have personally benefited from virtually all of them. Certainly, I can tell you that I have first-hand knowledge that injections, urethral suppositories, vacuum therapy and Viagra all work. My wife and I have restored a vibrant relationship. Our ability for sexual expression continues, unhampered by the toll that prostate surgery took on my erectile functioning. I awake each morning with gratitude that I have overcome prostate cancer, and that I live in such a time that my need for sexual expression can be so wonderfully met.

I came through the surgery extraordinarily well. Even my doctor was impressed. I give credit to the health habits I had already established and my use of principles, recommendations and products of the alternative health movement—filtered through the prism of alert common sense, of course. For example, I had taken saw palmetto for years—a long-recommended practice of alternative medicine people. In my case, the good outcomes many would expect from such practice occurred—my prostate gland was not enlarged and it functioned very well. Moreover, the cancer was a low grade when detected and was still contained entirely within the small walnut-shaped gland. Of course, having a great urological surgeon didn't hurt either.

From Helping Myself To Helping Others

As you can see, my original motive, which was to help myself, has expanded. I now want to help every man who may have an ED problem, as well as to assist all men who wish to maintain their normal erectile functioning. It pleases me enormously to consider all the marriages that could be healed and revitalized through application of the information presented here. I hope it has a far-reaching positive effect on the children of these couples and their lineage.

Many would agree with me that men's erectile functioning or lack of it, to whatever degree, influences all areas of their lives. To put it bluntly, men, by and large, identify—at least to some degree—with their penises. So when their penis changes, their perception of their world changes. For example, a man's self-confidence can be significantly bolstered by restoration of his erectile functioning and vice-versa. This spills over into productivity, relationships, attitude and many other areas. Given the self-confidence aspect alone, it becomes obvious that a lot of people can be affected— even those remotely associated with the man in question.

"An Ounce Of Prevention Is Worth A Pound Of Cure"

We're talking about a more serious issue here than what you may have thought. It is not just in the best interest of the men who are beset by erectile functioning difficulties to absorb and apply the information available in overcoming them. It behooves all of us, both men and women, to better inform ourselves. We need to also support and encourage those whose functioning is normal to do the things recommended in this book so they can avoid the onset of such dysfunctioning. To develop and maintain excellent health habits as soon as possible is the key. As youth gives way to the aging process, our standards of male sexual functioning need not be compromised. You can be vibrant and healthy and fully functioning well into old age, indeed, to the full extent of your life.

In a very real sense, this is a "self-help" book. Self-help is an honorable tradition in many other arenas and can be effective in enhancing your sex life, as well. If you wish to improve your ability to engage in and enjoy sexual intercourse, this book can help you.

You need not feel embarrassed or guilty about having such a goal. You are simply responding to a deep and natural human need. After all, sex is among the strongest of human drives and is what enables us to procreate. Sexual awareness is with us every waking hour and intrudes persistently into our sleeping ones as well. Sexual thoughts continually pass through our minds throughout the day and pervade our dreams as we sleep. We cannot deny our sexual nature, and moreover, we need not. It is ours to manage responsibly so we can fully enjoy the pleasure and fulfillment it can bring.

What This Book Can Do For You

This book gives you useful information about various approaches for overcoming ED. It also covers advances in medical science that make it possible for many men who've been experiencing problems in this area, to have a strong "hard" erection, whenever they want to.

The information in this book is for men who aren't satisfied with their sexual performance or, perhaps, aren't even aware that they're performing below their true potential. The information presented will enable them to enhance their sexual self-expression and the quality of their lives. It can help them enter confidently into a romantic situation knowing, without a doubt, that they will not be hampered by erection problems. This knowledge alone, can enhance the occasion by relieving anxiety and clearing the way for sexual expression.

Six Basic Approaches For Improving Erections Are Identified and Discussed, Including...

- **Physical Fitness**—Improving the physical condition of your body, especially those parts of the body that have a direct influence on your ability to achieve an erection.
- **Nutritional and Health or Medicinal (Non-Prescription) Supplements**—Most men are looking for something they can eat or swallow that will magically cause them to get a strong erection. Remarkably, there are such aphrodisiac substances! Yohimbe is such a substance. (See Chapter 4.) Some actually work for some men. Some have the approval of medical doctors and work for most men.
- **Pharmacological**—There are some remarkable pharmaceutical substances that can give a man an immediate erection. *Yes, it's true—a firm and ready erection can be yours . . . almost on demand!* What this can mean to you in a romantic situation is obvious. You will be delighted and your mate is likely to be appreciative—beyond words. Pharmacological also includes the introduction of what, throughout recorded history, has always been the most sought after approach to overcoming ED—an ingestible substance that works effectively. Sildenafil citrate, available from Pfizer as Viagra, is that revolutionary ingestible substance.
- **Vacuum Therapy**—This is an easy-to-use therapy that almost any man can administer to himself to produce instantaneous results. It requires the manipulation of a specially designed vacuum device but, the result is well worth the effort—and your lady will be pleased.

- **Conditioning the Penis**—This is an effective conditioning therapy which you can use on an ongoing basis, as will be described. Some men who have been sexually inactive for a period of time find that just using this self-administered "cold/hot" treatment, alone, brings them back to full potential. Others find that routine use of it can be depended upon to produce a very satisfactory erection.
- **State-of-Mind**—Much of the mechanics of getting and maintaining an erection is influenced by the mind. This includes your moods, emotional state, and the deep down beliefs and attitudes which are entrenched in your subconscious mind. This book discusses what you can do to achieve a healthy state-of-mind and a belief system so you can free your body and your sexuality in functioning or performing to your full potential. This introduction, to what may be called psychological therapies, will give you the foundation to explore your need, if any, for this kind of therapy.

It is my hope that the information in this book will enable any man to have better erections, if he wants or needs to, so he can experience restored or enhanced performance and all its benefits!

"... Viagra doesn't have the power to make or break a marriage. It can only facilitate increased blood flow into the penis—provided the man is sexually aroused. Viagra is like any other 'tool' used in life, in that what it is used for depends entirely upon the user..."

Chapter 1

AT LAST—POTENCY FROM A PILL... VIAGRA TO THE RESCUE!

"A dream for many men becomes a reality."

It's Here At Last...

Prior to its release, urologists had been telling their patients, for months, that there would soon be a pill for curing impotence. Imagine a man hearing this kind of encouragement just as he is about to undergo the biggest threat imaginable to his erectile functioning—a complete excision of his prostate gland. He'd probably be doubtful and think, "Yeah, sure." But incredulous as it may have sounded before its release, as you most likely know, that pill is, in fact, now here. Fortunately, the rumors of an impending medical breakthrough in the treatment of ED were not just wishful thinking after all. Vince, a friend of mine with ED, called me, way back in August 1997, asking, "Did you see what they were saying on television?" He went on to explain that he had heard a discussion on ABC television's "Good Morning, America" about a pill that "cures impotence."

On November 17, 1997, the news formally broke: *a pill that promises to cure impotence in 15 to 20 minutes*[2]. This pill promises to fit perfectly into the natural sequence of events in achieving an erection. If you are in a sexual situation and become sexually stimulated, Viagra will enable 70% of all men to achieve an erection. The other 30% need to discover an alternative that works for them, like the other solutions presented in this book.

Still another pharmaceutical company is reported to be developing a pill that enables an almost immediate erection. It dissolves under the tongue, and is said to be effective in about 70% of men whose ED stems from psychological causes (see Chapter 10 on the "Mind Factor"). The drug used as the basis of this second pill is reportedly a version of apomorphine. It is not expected to be available until sometime in 1999, but the interest and demand for this product is strong and the manufacturer will surely strive to have it available as soon as possible.

The oral pharmaceutical featured in *Newsweek* (November 17, 1997) is the closest thing to the "aphrodisiac-like" substance that many men have wished for—for years, and perhaps even centuries! This drug, sildenafil citrate, is being marketed under the trade name Viagra. It is manufactured by the well-known drug and pharmaceutical company, Pfizer. This "blue film-coated rounded-diamond-shaped tablet," as Pfizer describes it on page one of their website, received FDA approval on in late March 1998 amid deservedly great fanfare from all aspects of the media. Strictly speaking, it is not an aphrodisiac since it doesn't induce desire. But it does what that mythical pill, which men have been seeking since time immemorial, is supposed to do—*it enables most of them to have an erection.* The news about this latest triumph over ED is outstanding!

The most serious concern is about people with any heart disease or related condition like high or low blood pressure—especially older people with these conditions and those on nitrate medications. We'll discuss this a little later in this chapter. In all cases, no matter what medication you're on, consult with your doctor about potential problems with combining medications. Your physician may also be concerned about side effects, depending on your current state of health.

The side effects of Viagra are minor and, according to reports, they affected only a small number of the men involved in the clinical trials. The side effects reported so far are said to include flushing, moderate to severe headaches, some stomach upsets, body aches, nasal congestion, urinary tract infection, diarrhea, dizziness, rash and some mild visual disturbance, i.e., color blindness that could affect up to three percent of users who may have trouble distinguishing blue and/or green. But all these side effects, in the few individuals who experienced them, were considered to be mild and transitory. Talk with your physician about any other side effects he is aware of. If your doctor prescribes Viagra for you, your pharmacist is to provide you with a patient information sheet which lists side effects, precautions and so forth. Be sure to ask for it if it isn't given to you. And also be sure to read it!

The risk of side effects has not at all diminished interest in Viagra. Speaking of them, Dr. Harin Padma-Nathan of the Male Clinic in Santa Monica, California and a member of the Department of Urology at University of Southern California, observed that given the large number of Viagra users, and the fact that most have existing health problems, the side effects have been minimal. Dr. Padma-Nathan may be among the most knowledgeable people about Viagra as he was among the doctors who conducted the clinical trials.

Side effects or not, the effectiveness of this drug, especially for men like Vince, who are looking for something that could give them more consistency in their erections, has been remarkably high. Men who, up until now, have had inconsistent erections are experiencing the reliable potency. This indicates a great boon for a vast spectrum of men whose libido is intact, but who are getting less of a response in their penis than they would like. Different doses of the active ingredient in this pill have been tried, and it has been found that for some men only a small dose is required. You might expect that the larger the dose, the more effective it is in all men who can be successfully treated with Viagra—but that is not always the case. In fact, a man requiring only the smaller dose will not experience a significant improvement by taking a larger dose. Your doctor can discuss dosage with you, as appropriate.

So Should You Take Viagra?

This is an important consideration because taking Viagra is not without some risks. This precaution is particularly directed to any man (or woman) with any heart disease or a related condition such as high or low blood pressure. And this precaution is doubly directed to older people with these conditions.

Viagra has been assessed by doctors to be remarkably safe but this should not lull you, as a prospective user, into a naïve sense of complacency. Actually a caution from me is redundant because a medical doctor stands between you and the obtaining of the medication.

If you even have a fleeting thought of borrowing Viagra pills from a friend or a neighbor, I suggest you forget it. *Impotence, to any degree, is a serious medical condition and Viagra is a prescription drug, to be taken only under your physician's supervision.* Your doctor is obligated to determine whether or not you are a fit candidate to take Viagra. Here are some words of caution:

> You need to truthfully reveal your health history to your physician so he or she can make an accurate assessment of your health situation to take Viagra, or any other medication for that matter. It is particularly important to be truthful if you are being treated or have ever been treated for any heart, circulatory or blood pressure condition. Discuss with your physician any conditions you have for which you are under another doctor's care and any medications they have prescribed for you. Take all your prescription bottles in with you when you see your doctor, especially if you are under more than one doctor's care and are taking several medications. If your doctor prescribes Viagra for you and you receive medications from more than one pharmacy, be sure to take all your prescription bottles in to show your pharmacist, just to be on the safe side.

The few deaths that occurred among users during the early weeks of the availability of Viagra to the general public has caused some concern. However, the small number of deaths don't appear to be statistically significant. They occurred among an older and less healthy population who were already at risk. But since most of the deaths occurred among older men with previous heart or circulatory problems, speculation persists as to whether or not Viagra played a role in their demise.

Pfizer has specific warnings to patients being treated for heart conditions, like angina, for example: if you are being treated with nitrates in any form, (like nitroglycerine), you are not a candidate to use Viagra. Combined with Viagra, nitrates could cause a drop in blood pressure, resulting in fainting and even death in some cases. You must report any such condition to your doctor, and discuss it thoroughly with them.

Another concern arises, especially for older men who have not performed sexually for a long time, particularly those who are sedentary and out-of-shape in general. If this describes you, consider whether or not a return to vigorous sexual activity could, in any way, be too much of a shock for your heart and perhaps cause a heart attack.

Again, it's important to stay in shape at any age. The return of your potency must be handled wisely. So with your renewed potency, you must be honest with yourself.

Ask yourself, "What kind of shape am I in? Is my heart strong enough to engage in renewed sexual activity?" These are also key questions to ask your doctor, especially if you're even the least bit unsure of whether your heart can take the extra strain.

Men in the age group of 65 or older who have otherwise maintained enough vigor to be interested in Viagra are certainly a hardy group—at least in character and spirit. They are slow to complain or admit frailty, but in this case, they must level with their doctor. If the doctor advises against Viagra, there are other solutions for overcoming ED, which are covered within these pages. Viagra is surely a "miracle" drug for most men. However, if you are among those for whom it is not, in consultation with your doctor, explore the other solutions. To those of you with heart or circulatory problems who absolutely must not take Viagra, I suggest you consider the noninvasive vacuum therapy and check it out with your doctor. *So, should you take Viagra? This is a question you need to discuss honestly and thoroughly with your doctor so he or she can make a proper diagnosis and recommendations, based on the facts.*

If Your Doctor Decides You Are A Good Candidate For Viagra–Then What?

OK, lets's say Viagra will work for you and neither your physical or your health condition is an issue. Even so, should you still take Viagra? There is yet another aspect of taking Viagra that you need to consider. To reap the full benefit of this pill, you need to be in a relationship. Viagra, by itself, will cause no sexual sensations in you or make you more desirous, desirable or sexy. In fact, the decision to take it seems most appropriately the mutual decision of a man and a woman in a relationship. The fact that Viagra is effective only in a sexually stimulating situation suggests that it could be an ideal tool for enhancing the marriage of a couple dealing with ED. It acts by facilitating the man's ability to have an erection, which enables the couple to more fully express their love for each other. A fulfilled life including sexual fulfillment, is your birth right. Your reading this book indicates you want to experience this. Viagra can be a wonderous means of overcoming an obstacle to sexual fulfillment, but its effectiveness is best experienced with a loving partner.

Curiously, as reported in the media, Viagra was originally being developed as a heart medication. When the researchers decided to end the study, because they were not achieving the results they hoped for, they noticed that the men involved were reluctant to give up their pills. They reported having better sex, which they hadn't experienced before! When the researchers discovered this, they must have soon realized that they were on to something big.

Blood flow is a crucial factor in achieving an erection. In injection therapy, as you'll read about later, a pharmaceutical that ultimately increases blood flow is injected directly into the penis. Once injected, this substance, usually a form of prostaglandin, relaxes the smooth muscle vascular tissue which increases blood flow to the penis and ultimately results in an erection. But this new drug, sildenafil citrate (the active ingredient in Viagra) is able to relax this same smooth muscle tissue by being generally present in the body as a result of taking the oral dosages. As long as it is present, i.e., it hasn't worn off, its potential to produce an erection remains.

In injection therapy, as you'll learn, an erection will occur almost immediately. With Viagra, an erection occurs only in response to sexual stimulation. This indicates that in some forms of ED, it might not be effective. An example of such a case would be that in which damage had occurred in those nerves through which the response signal from the brain travels to the penis. Any man who has undergone prostate surgery has run the risk of having had these nerves damaged, or worse, cut and sections removed. In such a case, whether the oral medication would work or not is problematical. However, vacuum therapy and injection medication would still be predicted to work very effectively. The best advice would be to take the Viagra pill and find out, if, of course, your doctor will concur, and give you a prescription.

Just the same, for most men, this pharmaceutical sounds like a gift from heaven and demand for them is and, in all likelihood, will continue to be exceedingly high. For those men whose erection difficulties are treatable by this kind of therapy, the promise is that once they have taken the pill in the prescribed dosage and manner (perhaps about 30 to 60 minutes in advance of intercourse) they will have an erection in a sexually stimulating situation.

Just in case you're wondering about that second pill I mentioned earlier (expected to be available in 1999) or whether there are other pharmaceutical companies developing their own or similar oral ED treatment drugs, the answer is "yes." For example, the pharmaceutical company Zonagen has a pill (Vasomex) that has also undergone clinical trials. It is reported to be safe and it works, though it must be added that clinical test results were not as dramatically positive as had been hoped or expected. Just the same, they were convincing enough to send the valuation of Zonagen stock climbing, according to reports, to more than $500 million. Vasomex, too, enables a man to achieve an erection in a natural manner when sexually stimulated. Vasomex contains phentolamine, a drug already approved for treating high blood pressure. As with sildenafil citrate, an erection is achieved when the man is sexually stimulated. Interestingly, even though Vasomex is available in Mexico by prescription, Mexican men were nevertheless caught-up and consumed by the Viagra craze that swept in from the United States following its introduction.

Is a better pill on the horizon? Icos Corporation of Bothell, Washington is working on it and hopes to have a rival pill available sometime in 2000. The pill doesn't have a name yet, but claims are being made for it that it will have fewer side effects. Apparently, so they say, this is because it is more narrowly focused chemically, acting more narrowly on the enzymes that influence relaxation of the smooth muscle tissue that enable erections.

A Word Of Warning!

With the advent of Viagra, an interesting question faces the large number of men, like Vince, who just want to "get it harder." Should they drop those more natural substances that may be working for them for the more potent chemical solution presented by these pharmaceutical ED cures? Some physicians and healthcare professionals have expressed this kind of concern. Their reasoning goes something like this – *ED is symptomatic of more serious underlying conditions.*

We know there are many cases of doctors examining men only to discover that the cause of their ED is actually heart disease. The small arteries that supply blood to the penis get narrowed by fatty deposits the same as do the larger ones that supply blood to the heart! *A man will have direct experience that blood flow to his penis is not what it should be, well before the condition worsens and a heart attack tells him that blood flow to his heart is restricted too!* Therefore, taking a pill to increase blood flow to the penis may result in a patient ignoring a more serious condition.

This reasoning is certainly sound and I encourage you to see your sexual functioning as part of overall healthy body functioning. Remember that exercise, a proper diet, and the resulting good blood flow, are the keys to good erectile function. I would also remind you of the salient message of the alternative medicine movement *that natural herbs and remedies pose much less risk to the overall well being of the body than the potent chemicals resulting from medical research.*

Vince's response to this caution is, "This is the miracle drug we've been waiting for; don't knock it." Perhaps he is right. Viagra is here now. On March 20, 1998, much of the western hemisphere saw Dr. Timothy Johnson on the ABC Television's *20/20* announce to the world that the FDA had at last approved a pill, Viagra, that worked for treating male ED. Viagra blocks a certain enzyme that is found almost exclusively in the penis and, when this enzyme is blocked, it permits an erection to occur. As it turns out, this drug works mainly in the penis, and the best news is that it apparently has little effect in any other part of the body! At the present time, the news is tremendous, and there seems little reason for concern. But, as in all things, keep informed and exercise good judgment and caution.

Obtaining And Using Sildenafil Citrate (Viagra)

The pill of choice is, of course, Pfizer's Viagra. Viagra is a trademark and no other manufacturer is likely to be able to offer sildenafil citrate in the foreseeable future. So beware of imitation and fakes and frauds. They are sure to appear. Viagra can only legally be obtained by prescription. It's been reported that a number of illegal sales are going on in countries where Viagra hasn't been approved yet. The retail price for one pill, which will probably continue to rival the cost of a bottle of moderately fine wine, is around ten U.S. dollars. Reportedly, some pharmacies, especially those in discount stores, are offering the pills for less, in an attempt to attract more business.

For some, Viagra will be covered by their health plans. Initially, one HMO (Health Maintenance Organization) in the U.S. has announced that it will allow up to six pills per month, once a need for them has been verified. Other HMOs are setting rather strict guidelines as to who will be able to obtain coverage. There is, as might be expected, disagreement on what to include in the guidelines. Another HMO, for example, is said to require testosterone testing as an initial indicator of ED, even though it is known that some men with perfectly normal levels of testosterone may have ED from some other cause.

During the first few weeks following the introduction of this pharmaceutical, the relatively high price did not seem to be a concern—the demand for the pill was high and it continues to be high! Men like Vince were "coming out of the woodwork." Vince, himself, had managed to get a prescription and phoned to tell me about it.

"What?" I ask, pretending to be surprised. "You don't really need this kind of medicine, do you?"

"This stuff works," he said. "They gave me the smallest dose 25 milligrams."

"What did it cost you?"

"Cost doesn't matter," he replied. "It's worth it! You can think of it as a part of the cost of an evening out with your wife. And if you don't go out—that makes the cost of the pill seem even more reasonable!"

I had to admit that I was already a Viagra convert. I had left a telephone message for my doctor, while it was still in clinical trials, requesting a prescription. When he returned my call to tell me that it was not yet available for prescribing, I wasn't home and he talked with my wife, who responded with some incredulity, "Why does he want that? He's doing just fine with what he's using now."

I was amused when I learned of her reply—and I was pleased too. Just the same, I was very interested in this pill. I had even made a half-hearted attempt to participate in the clinical trials, though I did not earnestly pursue it.

Immediately after Viagra was approved, I may have been one of the very first in line to obtain it. And I wasn't disappointed! My doctor wasn't sure about what strength to prescribe for me. On the one hand, I had undergone a radical prostatectomy, but on the other I required only modest strength using injection therapy. Also I was able to use vacuum therapy very successfully. In the end he decided to prescribe the maximum 100-milligram with the instruction that I experiment with just half a pill and increase to a whole one if I needed to. On my third use, I opted for a whole pill and decided that it was the right dose for me.

I used all of the solutions I just mentioned and they all worked well for me. On the spur of the moment, my wife can signal me, and I can leave the bed momentarily to apply vacuum therapy, for example, and be back quite prepared...in less than ten minutes. But on the other hand, with Viagra, its much like it use to be—with everything seeming to flow naturally from one step to the next. Also I did not experience any of the reported small Viagra side effects, even when I was taking the maximum strength.

Vince had received the pill in the smallest strength. It comes in 25, 50, and 100 milligrams; but the pills cost the same whatever strength you get. Not surprisingly, some men are asking for the higher strengths and cutting them in half. It is not clear whether or not this is an advisable practice or not, as there is some question as to whether or not this affects the absorption rate in a manner that will lessen the pill's effectiveness. From my own experience, I don't believe that it does.

In Vince's case, he experienced no side effects. But I know of an older man who did experience a headache after the first use, though he was very pleased with the resulting erection he experienced. The worst report came from a 65-year-old man who not only failed to get an erection, but also got a headache. His result was not unexpected as he had previously undergone radical prostate surgery. Those vital nerves that connect his penis to his brain had been damaged.

The biggest expected market for Viagra had presumably been

older men who would thereby be given increased control of the later third of their life. But younger ones (35 to 45 and even younger) are going after it in droves! Indeed some medical clinics immediately recognized that a need existed in this group. By the fourth week after the Viagra introduction, advertisements appeared in major newspapers featuring a sculptured young male body bare from the waist up, advising the reader to "Be the man you used to be—Viagra is here now."

Viagra, Society and Personal Responsibility

One more obvious result of Viagra's availability is that it has brought impotence "out of the closet." It's now a common topic of conversation. Many men who have ED, perhaps for years, are now more comfortable admitting it and are getting the assistance they need. People, both famous and unknown, are appearing on radio and television, admitting they have ED and sharing they're now using Viagra or some other solution. For example, former U.S. Senator Bob Dole of Kansas, a survivor of prostate cancer surgery, had participated in the Viagra clinical trials. In fact, Mrs. Dole reported that she and the former senator are quite pleased with Viagra.

Generally speaking, men taking Viagra are reporting better sex including more intense, explosive orgasms, more frequency, and of course, as expected, firmer, stronger erections. Sociologists are speculating on what impact this marvelous medication is going to have on society. When Viagra made the editorial page of the *Los Angeles Times*—the article indicated that the impact of this pharmaceutical has been even greater than anyone expected. In fact, as of June 1998 over a million prescriptions had already been written!

Some of the early complaints against Pfizer and Viagra were exemplified by a court action. A woman blamed Viagra for causing her husband to leave her after he regained his potency through using it. While regaining potency may lead to actions that make it clear a marriage is troubled, Viagra, itself, doesn't have the power to make or break it. It can only facilitate increased blood flow into the penis—provided the man is sexually aroused. Viagra is like any other "tool" in life, in that how and what it is used for depends entirely upon the user's level of personal responsibility.

For example, the newly sexually empowered male may finally realize, or have it confirmed, that his marriage has been faltering. His newfound ability to have an erection, just made it obvious. The truth is that the origin of his discontent, and most likely his wife's, preceded his taking Viagra. To blame Viagra, or anything or anybody else, would be irresponsible and not facing the truth. Viagra, itself, cannot cause unhappiness or unfaithfulness or the results stemming from either.

When it comes to using Viagra, it's the man's mind, not Viagra, that chooses how he regards his marriage. If he chooses to be unfaithful to his wife, that's unfortunate. But, he will face the consequences of his action, just like anything else he chooses to do in life. In reality, the mind is the ultimate sex organ (see Chapter 10—The "Mind Factor").

Viagra needs to be used responsibly and considerately to enhance a marriage relationship. While it is certainly exciting to be able to have sex again, talk with your wife about it first, and be careful not to wear her out. Be considerate of her needs.

A successful marriage depends on how much the husband and wife want it to work. If the relationship is troubled, they may want to seek counseling *before* using Viagra.

It is my hope, that with this book, the ability to have renewed intimacy because of taking Viagra, or finding another solution to ED, will lead to greater bonds between husbands and wives.

How Could A Meal Affect Your Success With Viagra?

Say you have just received your prescription for Viagra and you're excited. You and your wife decide to go out to your favorite restaurant and celebrate with a big, romantic meal. As you finish your dessert, you swallow a Viagra pill so you're ready when you get home an hour later. During the drive home you are both full of anticipation of renewed intimacy.

Finally you're at home, in your dimly lit bedroom, holding and caressing each other, getting hotter and hotter. What a disappointment when you don't get an erection, even a half-hour later. What happened? You may be thinking, "Well, I'm one of the men Viagra won't work for." Don't be discouraged. You may have eaten too much, especially high fat foods, and your digestive tract couldn't absorb the Viagra.

On the opposite end of the spectrum, some men can have an erection in as little as fifteen to twenty minutes, on an empty stomach! So, if you ate too much first, don't rule yourself out as a good candidate for Viagra. Next time, just eat a light, low-fat meal before you take Viagra and watch what happens!

If Viagra still doesn't work for you, there are other solutions in this book you could use. Consult with your physician, keep an open mind and, unless you have a serious health condition, it's likely you'll be able to overcome ED.

Viagra...For Women?

Since Viagra works by increasing blood flow in the pelvic area, it could also have application to women, as sexual function in them is crucially connected with blood flow. The question of whether it could help women was almost an afterthought, but in fact, indications are that it does. It has not yet been approved by the FDA for use by women, but clinical trials to determine its applicability for them are being conducted.

A Women's "Orgasm Pill" Is On The Horizon!

Viagra is not the only pill being looked at as a possible enhancement to women's sexual functioning. It's probably only a matter of time before an "orgasm pill" for women will gain worldwide attention. On June 7, 1998, a Reuters newswire from London revealed to the public what many medical insiders already knew—that such a pill is being tested in clinical trials in Europe. This women's pill was reported to be "impressively effective" in clinical tests involving 500 women across Europe. Although Pfizer's Viagra is being tested in clinical trials for women in the U.S., as noted before, Pfizer is also reported to be in the running to market this new pill.

Chapter 2

Yes – You Can Be More Potent…Even If Viagra Isn't For You!

"Most people aren't aware of how many solutions there are to ED, in addition to Viagra."

A Man's Search For Answers

Larry presents a most enlightening case of a man intent on helping himself overcome ED. I feel it's important to share this story with you since Viagra won't work for 30% of all men! Also, many others may not even be able to take it because of other medications they may be on, their health, religion or other beliefs they may have.

Although I had known Larry for a number of years, it was only when he cautiously took advantage of an opportunity to bring up the subject that I learned of his predicament. It happened late one afternoon after we had finished a round of golf, and he accompanied me to pickup my wife, from her work as a registered nurse in a medical claims review service. Neither of us could ignore that she was finishing up a telephone call about a rather interesting topic. We couldn't help but notice that she seemed to be talking about the denial of insurance payment for a penile implant.

"Penile implant?" Larry whispered, "What in the world is *that*?"

Later, when we were alone, he brought up the subject again. I soon realized that Larry had taken this opportunity to bring up a subject about which he was both very interested and at the same time, like most men, embarrassed about. Fortunately, he overcame his embarrassment and soon got right to the point.

"You had prostate surgery," he said, "yet you and your wife still carry on like newlyweds! I heard that operation is supposed to take the 'gunpowder' right out of you. What's your secret?"

I didn't know it then, but this was the beginning of my compiling all the information I had gained or learned to inform men like Larry. Larry was my inspiration. He became my catalyst to organize and expand the knowledge I had gathered to overcome the challenge of ED that prostate surgery had presented to me.

Larry asked me rather sheepishly, "You don't have one of those 'penile implants,' do you?"

"No, Larry," I said, "fortunately I don't need one of those. It's great that they have them, though, for the guys who really need them."

It was when Larry asked the first question again that really got my attention—"What's a penile implant, anyway?" he asked.

I was first amused by his lack of knowledge. It then occurred to me that he was honestly uninformed and that there must be many other men who also lack information about dealing with ED. Thinking back on it, I had been one of them myself as I faced prostate surgery and the prospect of permanent ED.

I explained that a penile implant is a device, which is surgically implanted directly into the penis, to enable a man to perform sexual intercourse.

His reaction suggested shock and even disbelief. I was surprised that he had not heard of penile implants. So I gave him a brief synopsis of what I knew: "Penile implants," I told him, "are, in my opinion, a last-resort cure for impotence. There are several kinds. One of the simplest and most common is one that gives a man a permanent erection! Others require a mechanism to inflate the penis to erection—to literally pump it up. Some men, who were previously impotent, report an almost smug sense of confidence once they acquire an implant. And you may be surprised to know that they are often paid for through health insurance coverage." Nevertheless, like many men, Larry recoiled at the idea of surgically implanting a device into his penis.

"Isn't there something less drastic," he wanted to know "for guys who just want to get it harder?"

I recognized his point immediately. It was a question similar to the one I had to answer for myself as I faced radical prostate surgery. The answer to Larry's question is a definitive "yes"! As we talked, I realized there were probably millions of men who may need information on what options were available to get better erections. Or how about men with no obvious health problems but who, just the same, are consistently not having really satisfactory erections? And what about others who, from time to time, are not having satisfactory erections?

Based on the research I've done, and the stories I've gathered, with the assistance of Dr. Bolton, on the experiences of dozens of men, and most of all, from my own experience, I have found that the vast majority of men can have firmer, harder erections. In most cases, what you or any man needs, to have a better erection, is the right information. It is the purpose of this book to give you that information. Even if you have a serious health problem, depending on what it is, and how dedicated you are to restoring and maintaining healthy erectile functioning, you may be able to do so. Talk with your physician and share the information in this book with him. He may want you to consult with an urologist.

For a period of over two years, I followed Larry's experience as he dealt with his problem. Both he and I learned a lot which will be beneficial to most men. At that time, Larry was forty-five years old, looked to be in good health and seemed to be fairly fit, even though he did not work at physical fitness. He got married at nineteen, but was divorced before his twenty-first birthday, while he was away serving in the military. Later he returned home and went to college. After graduating, he got a job with a large retailing company and worked his way up to a managerial position, where he is today.

This book is the answer to Larry's question: "Isn't there something for guys like me who just want to get it harder?" It also answers the question, "I have total ED: what can I do to restore my potency?" This information applies to men of all ages, but is especially helpful to men thirty-five to seventy-five, and beyond, if they are physically active.

By the time I met Larry, he was just coming out of a second failed marriage. I soon learned that he had been shattered by its breakup. When he honestly attempted to face the reason for it, he had to admit he wasn't sure whether it had anything to do with what he described as "his weak erections." "Sure," as he described it, "there were times, I had to 'stuff' it in." But he could remember only once that his wife had made any comment about it. "Why is it so soft?" she asked. Amazingly, he was able to ignore her question and only thought of it after the divorce. Then he was crushed.

Larry had to admit he had been unwilling to face and deal with his ED. Maybe she would have been supportive and patient. Of course, now he'd never know what her reaction would have been had he been courageous enough to admit he had a problem. It looked for awhile like he would never have a successful marriage, that is, unless he got some help.

Love Strikes Again!

Since he felt basically healthy, Larry's libido was intact and he was an otherwise well-adjusted person, it was not very long before he found himself growing ever more involved with another attractive woman. After an evening with a candlelit dinner and rather romantic dancing, he knew it wouldn't be long before he would want to be married again. Thus began his frantic search for a way to be assured that he would be able to have a strong hard erection and a solid sexually exciting marriage. He read everything he could find, searching everywhere he could think of.

Like most guys, he started by looking for things he could take orally such as foods, vitamins, other nutritional supplements, and herbs, roots and potions synthesized from them. When his wedding night came, despite his preparation, things didn't go all that well. He would soon learn that vitamin E and oysters won't help a lot if you are full of anxiety! More importantly, he got excited when he learned that, with his seemingly good health and sexual desire intact, an easy-to-use pharmaceutical would have guaranteed him success.

Luckily for him, his wife is a passionate, caring woman who seemed to ignore his poor performance. When he had breakfast with her the next morning, it was obvious she was genuinely glad to be with him. That was a relief. But, he thought, "I need to find a solution that works for me *this* time. I'm going to see my doctor and ask for a prescription for Viagra."

When he arrived for his doctor's appointment his blood pressure was checked, along with some other routine tests. When he met his doctor, Larry explained his ED problems and asked for a prescription for Viagra. Shockingly, the doctor said, "No! Larry, your blood pressure is too high—first we have to deal with that and the medication I am going to put you on should not be taken with Viagra. For now, we'll resume blood work and other tests to find out what's going on. In the meantime, I can prescribe vacuum therapy. When we get your blood pressure under control, we can take another look at Viagra."

Overcoming Impotence And Enhancing Erections

This book contains detailed information about the latest advances available for overcoming impotence and enhancing erections. Medical science has made what some people consider miraculous advances in treating erection problems. The time has come for better, more convenient and easy-to-use treatments.

The promise of these advances is that you now have, virtually at your fingertips, a variety of possible solutions that can give you the ability to achieve a stronger, firmer erection. There's a great likelihood that you, too, will be successful in your own "quest for potency." In fact, as was true for me, you may find several solutions that work for you as well.

In the next two chapters, we will be looking at some of the basics of a man's sexual functioning—namely, blood flow, exercise and that all-important hormone—testosterone. You will learn about nutritional substances, especially those that have an aphrodisiac effect, and about therapies that restore potency. From there we will go to a variety of topics on enhancing erections. Most importantly, you will learn about medical breakthroughs that can literally give you "potency-on-demand." This is exciting news for men with any degree of ED.

Whether the problem is "getting-it-up" at all, or getting a firmer erection, there are medical solutions that can now be applied to every degree of ED. They include discoveries about substances similar or identical to those produced in the body that relax vascular tissue in the penis, inducing an immediate erection. There's also physical therapy that can do the same thing, as well as an array of other treatments. In short, I believe there is a solution for every man.

Pharmaceuticals, Other Than Viagra, That Can Help

The pharmaceuticals behind these extraordinary treatments are prostaglandin, papaverine, and phentolamine. When they are introduced into the penis they mimic the natural mechanisms of sexual stimulation. The blood vessels relax, blood flows freely into the spongy erection chambers inside the penis, pinching off the blood flow out and, *voila*, the penis, now engorged with blood, is erect. When this treatment is applied in a sexually stimulating situation, the effect can be very gratifying. Using these substances, nearly any man can achieve a satisfactory erection in about ten minutes or less. The few exceptions would include men with severe penile injury or blood flow problems.

Pharmaceutical companies are continually improving ways to introduce these substances into the penis. When they were first made available, a small syringe was the only option. Now pharmaceutical companies, like Upjohn, have introduced a specialized device, sold in kits of several doses, that simplifies injection. Another company, Vivus, Inc., has introduced a method of placing the substance into the urethral canal where it is absorbed into the penis's two cylindrical-like spongy tubes called the corpus cavernosum, to produce the same results. And, wonder of wonders, a substance that can be taken orally to effectively produce an erection, in just minutes, is reportedly under development.

But as all the world knows, Viagra, the "potency pill," considered by many to be *the* "miracle pill" will most likely be available in pharmacies worldwide. It is an effective treatment for 70% of men with ED. Although 30% of ED sufferers may not be helped by Viagra, and still others won't or shouldn't take it, many still consider it to be exactly the substance men have been searching for since the beginning of time. But whether you are among the lucky 70% or the remaining 30%, the information in this book can help you.

A Medical Technology That Can Help

The pharmaceutical successes in overcoming ED are only part of the story. Medical technology has produced a very effective therapy that can enable almost any man to achieve a satisfactory erection as soon as it is applied. It's called vacuum therapy, as was offered to Larry. You may be among those who scoff at the seemingly naive idea of using a pump-like device to achieve an erection, but it is actually a very sophisticated application of the principles of vacuums and it is easy to do.

In the simplest explanation, a vacuum is created around the penis allowing it to become erect as it is engorged with blood in response to being in this vacuum. The flow of blood into the penis is in every way similar to the way an erection would be achieved naturally in sexual arousal, except that the increased blood flow was not achieved by the relaxation of vascular tissue. Many of the thousands who use it consider it almost a miraculous return of their erectile function.

Use of vacuum therapy and pharmaceuticals represent dramatic victories over ED. But there are other approaches to overcoming ED problems that are quite effective though less publicized. Mainly they involve the use of supplemental nutrients, but include some outstanding therapies as well. The explanation for the lack of publicity about them is probably due to the fact that knowledge of them has circulated mostly in the province of alternative medicine. Still, they will have application to many men, particularly to those who, like, "just want to get it harder."

A Word About Alternative Medicine

The medical establishment is increasingly coming to recognize the value and validity of treatments and products previously known and available mainly through health food stores, or from certain practitioners who rely on natural products rather than synthetic drugs. It is not uncommon now to see a prominent newspaper headline announcing that the medical establishment has recognized the effectiveness of yet another natural product.

A good example is a page two headline in the *Los Angeles Times* announcing "Herb May Help Slow Alzheimer's.[3]" The article reported results of a study that showed ginkgo biloba was as effective in slowing the progress of Alzheimer disease as other more conventional treatments. It would have been no surprise to those knowledgeable about health food products that 27% of patients given ginkgo biloba showed improvements in mental functioning. That's why millions of health food store customers have been taking it for years. One of the properties of ginkgo biloba is that it is a blood thinner and, like nearly every product of this kind, may not be suited for everybody. And yes, it does have application for improving erectile functioning.

Perhaps the best indication that alternative medicine is gaining recognition is the example of Dr. Andrew Weil, M.D. He is, perhaps, the most listened to and read medical doctor of recent times. His reasoned joining of alternative and establishment medicine has and is influencing millions. You many want to read his books and visit his web site.

Chapter 3

BASIC CONSIDERATIONS IN MALE SEXUAL PERFORMANCE

"Even if you are sedentary and you don't exercise much, there's still a lot you can do to improve."

Why Are Exercise and Fitness Important?

It's easier to have a strong erection when your blood circulation is good. Many men don't think they can do anything about it, but the good new is, they can.

Even a rather fit man who exercises regularly, may not be doing everything possible to maximize blood flow in the area of the penis. Furthermore, you don't need to train like an Olympic athlete to gain the benefits of better blood circulation. Any man who is at least moderately fit has a head start. And anyone who is not otherwise ill, and is willing to exert a little self-discipline, can improve his fitness and blood circulation.

The point is this, you *can* improve your fitness and cardiovascular health. The result will be improved blood flow to the penis for a stronger, firmer erection.

How Does An Erection Occur?

An erection is the result of blood flowing into the penis and held there by the action of the smooth muscles that surround the blood vessels. The penis itself contains no cartilage or bone. It's all a matter of blood flow to the penis.

Upon sexual arousal (meaning, getting "turned-on") a complicated series of events start to happen in your body. But ultimately, the brain sends a signal to increase blood flow into the penis while at the same time decreasing the flow of blood out. If everything is working according to plan, "getting-turned-on" automatically results in an erection. But, as we now know, it often doesn't work out that way for more than one-in-five men!

Blood is continuously flowing into the penis's *corpus cavernosum*. This spongy tissue contains smooth muscles, fibrous tissues, spaces, veins, and arteries. But under most conditions, the amount of blood that remains in these tubes is not enough to make an erection. This is where getting "turned-on" comes in.

Upon sexual arousal, the brain signal to the penis causes the two tubes to fill with blood resulting in a firm and erect penis. Here's how it works – the signal from the brain travels down the spinal cord and into the penis along a special set of nerves. Lying along the outside edges of the prostate, these carry the signal directly to the penis. There they connect with the arteries that bring blood into the penis, causing them to relax. (Yes, this is all a part of that complex response of getting "turned-on.") Then, an increased amount of blood flows into the two spongy tubes. At the same time, because the tubes are now engorged with blood, they press against the veins, which ordinarily let the blood out. This then closes the veins, almost like pinching a water hose, to hold back the flow. The blood is restricted from flowing out of the tubes, and the penis becomes erect. It then stays erect at least as long as you remain sexually aroused.

Relaxing of the arteries opens up the way for blood to flow freely into the penis. This flow raises the pressure inside the penis, blocking the veins that normally drain the blood out. As long as those arteries are relaxed, the penis remains stiff and erect.

There have been many discussions about nutritional substances or drugs that could aid in the relaxation of the arteries. One of the most promising speculations is that a lack of certain nutrients is a cause of poor artery relaxation.

This is promising because it suggests that a dietary supplement—taking choline, a B vitamin, and L-Arginine, an amino acid, possibly along with other B vitamins and amino acids—may help alleviate this deficiency, which is thought to be an absence of nitric oxide. (There's more discussion of nitric oxide throughout this book.) But even if this deficiency exists, the pharmaceuticals and vacuum therapy, as mentioned earlier can, in most cases, overcome it.

Remember, it's the brain that sends the signal that causes the arteries to relax. In this sense then, the brain is the ultimate sex organ! We will discuss getting "turned-on" in another part of this book under the topic of testosterone and the libido. But for now, let's continue the important topic of blood flow and what you can do to improve its circulation to the penis.

Exercise Is A Major Key

For most men, getting a erection in their 20's and younger is almost as natural and spontaneous as breathing. Certainly it was no problem. But most things having to do with the body were no problem in that stage of youthfulness. A man could eat almost anything he wanted without gaining weight and muscle tone was easy to maintain.

It begins to be a different story for many men over thirty when keeping fit requires some effort. It's about this time in life that men also start paying attention to their ability to get and maintain an erection.

Erections in the middle of the night, like miniature tent poles, may start becoming the exception rather than the rule. It's not surprising that, just as working out in the gym helps maintain muscle tone, it also has a beneficial effect on a man's ability to get and maintain an erection.

If you are physically fit, you are well ahead of the game. But if you are sedentary and you don't exercise very much, there's still a lot you can do to improve. Words of advice, if you are not as physically fit as you would like to be, seek the advice and approval of a physician before you start any physical fitness program!

What's A Minimum Physical Fitness Program?

Most healthy men can easily do a basic program of regular aerobic exercise, such as walking at least thirty minutes a day, five days a week. Let's say you go to a gym and consult a physical trainer about a minimum fitness program. You would probably be advised to do low-impact aerobics to build endurance without straining your body. This would probably entail three sessions a week of twenty-to-thirty minutes. Or, you might be told to engage in swimming, stair climbing, rowing, riding a stationery bike, or even running, for a similar amount of time on a weekly basis.

Ideally you can go to a gym and obtain the expert service of a physical trainer. But if for some reason you can't or won't go to a gym, I would strongly recommend that you involve yourself in a minimum physical fitness program of walking. This could be as simple as brisk walking, thirty-minutes-a-day, five-days-a-week. It will do wonders for anyone who previously has not exercised at all.

Amazingly, this modest amount of exercise will put you ahead of nearly everyone else, in terms of regular physical activity. Only about 50% of Americans, for instance, exercise on anything like a regular basis. When you pursue this program with diligence, you'll be walking the equivalent of nearly eight miles a week. Even at an average pace, you would be well on your way to reaching that adequate circulatory health so essential for a strong erection. You only need to add some strength and flexibility exercises to have a modestly adequate aerobic exercise program. The benefits of aerobic exercise are well-documented. The overall health gains and heightened sense of well-being that results from regular participation in an aerobic exercise program are significant.

What About A Physical Fitness Program For Optimum Results?

Fitness that promotes the best condition for attaining better erections, requires both aerobic and muscle development exercises such as using weights and resistance exercise machines. Two things happen to benefit your attaining better erections with these approaches. Aerobic exercises promote better blood flow and stimulate the body's production of testosterone. The muscle building exercises are necessary because they stimulate production of the receptors needed to make testosterone useful.

Testosterone is an essential ingredient in the proper functioning of the libido. It's the hormone that gives males their maleness. Research shows that the intensity of your exercise is key to gaining the greatest benefit in terms of increased levels of testosterone. What is meant by intensity? Here are some examples…

- Vigorous running, such as three to eight miles, at a pace of six to eight minutes per mile.
- Several quarter miles on a track at a pace of around two minutes per each quarter mile.
- Weightlifting—continuous repetitions of increasingly heavier weights, with only short rest times between sets.
- Thirty minutes of nonstop calisthenics, such as in an aerobic exercise class.

Without testosterone, the signal from the brain that produces an erection does not get delivered to the penis. With increased levels of testosterone, the potential for sexual arousal is increased. Fortunately, if your testosterone falls within normal levels, you basically have nothing to be concerned about. See your medical doctor to determine whether or not your testosterone level is low.

Your goal should be to attain the highest level of testosterone that is normal for you, and a vigorous exercise program can go a long way toward accomplishing this. In fact, even a modest program of walking thirty minutes a day would go a long way, for most men, toward maintaining fitness and consequently improving erections and increasing libido. An optimum program can have a profound impact on your entire life, increasing both your vitality and your self-esteem.

Intensity is the key for achieving the bodily condition that promotes optimum testosterone production. Achieving the necessary intensity is within the capability of nearly any reasonably healthy man. As the term aerobic exercise implies (aerobic means "with oxygen"), performing them requires an increase in the oxygen supplied to the heart.

Aerobic exercises are those that exercise the heart and the cardiovascular system. These systems cannot be exercised directly, but only through an increase in workload put on them by exercising the large muscle groups, mainly the legs. The goal in doing aerobic exercises is to attain what is called the "training effect." Doing this on a regular basis, at least five times a week for thirty minutes, is the single most important thing you can do to improve your health and vitality. And it will, without a doubt, improve erectile functioning for most men.

How Do You Attain The "Training Effect" And Verify It?

Your pulse rate is the gauge for determining the effectiveness of your aerobic exercise. A simple formula can guide you in determining this. To be precise, your exercise is effective to the extent that you attain the "training effect." What is the "training effect"? It's the point during exercise when the heart is required to exert nearly maximum effort to keep the body, particularly the large muscles, supplied with oxygenated blood.

What it takes to reach the "training effect" may be different for each person. The "training effect" is an individualized goal that each individual needs to determine for himself (or herself) depending on his or her age and physical condition. Attaining the "training effect" maximizes the effectiveness of exercising the heart and cardiovascular system. For that reason no two persons' "training effect" goals are necessarily the same.

Performing aerobic exercises to the extent that you attain and sustain the "training effect" can result in achieving a high level of conditioning.

Here are the steps you need to take to determine your individual "training effect" goal...

- Check your resting pulse rate either at your wrist or neck: count the number of beats for six seconds and add a zero. If you counted seven, your pulse would be 70. For a more accurate measurement, count the number of beats for fifteen seconds and multiply by 4. If you counted 16, your pulse would be 72. Or better yet, count your pulse beats for a full minute.
- Next, determine your "fitness factor." Here's how: *simply subtract your age from 220*. For example, if you are forty, your fitness factor would be 180. If you are fifty, it would be 170, and so forth.
- How fit do you think you are now? Decide what your present level of fitness is and express it as a percentage. For example, if you are totally out of shape, accept a training level of 60%. If you are in really great shape, try a training level of 80 or 85% or any number in between. Better leave 100% for later in your training program.
- Multiply the percentage number you have decided upon (based on your judgment of how fit you are) by the fitness factor. For example, if your fitness factor is 180 and your percentage is 70, meaning you are in fair or good condition, you will get "126" (70% of 180 is 126). This means that when you exercise, your intention will be to raise your pulse rate to 126 beats per minute. This is your individual "training effect." To condition your body you must raise your pulse rate to 126 beats per minute and keep it there for between 15 and 30 minutes. Do this at least four times a week. Some research

indicates that if you exercise in this manner five times a week, the benefit can be as much as four times greater.

- Be sure to consult with your doctor before starting any fitness program, or any change you may want to make in it.

What exercise could you do to attain the "training effect"? You have a variety to choose from. At the top of my list of suggestions would be a vigorous calisthenic program, which requires a high degree of determination and self-discipline. An easier approach is to use one of many exercise machines available, such as a stair climber, stationary walker, or stationary bicycle.

Remember that just plain running, jogging or bicycling are absolutely among the best aerobic exercises you can do. However, if you bicycle, you must keep in mind the necessity to attain the "training effect." Leisurely riding and coasting will not do it. An easy test for determining whether or not you have reached the "training effect" is if you can carry on a conversation? If you can, you probably are not in the "training effect."

As we will discuss later, to achieve the greatest benefit for improving erectile functioning, your exercise program should include strength-training exercises as well as aerobics or cardiovascular exercises. This, as we shall see, enables optimum testosterone production as well as preserves and develops muscles. We will look at other ways to increase testosterone level in the body in the next chapter.

How Can You Improve Your Blood Flow, Especially Where It's Really Needed?

Generally speaking, a good exercise program, like you might have under the supervision of a trainer in a fitness gym or workout club, could be ideal for the results you are seeking. This typically includes stretching, low-impact aerobic warm-up, circuit training on resistance machines, vigorous calisthenics, and weight training with barbells. But even the modest program of walking, as described earlier, can be of great benefit.

Giving attention to improving the blood flow to the pelvis and thighs, and adjacent muscles, can be particularly beneficial to improving your erectile functioning. And to that end, there are some specific exercises that can be especially useful.

As you progress in your own physical fitness program you will find others that may also be helpful to you, be open to exploring. Assuming you are moderately fit and not greatly overweight, the following two exercises, which are mainly muscle-strengthening, will improve the blood circulation in your pelvic, thigh and abdominal areas. Remember, these are just two particularly helpful exercises in improving your ability to have a strong erection. They are *not* substitutes for a full program of physical fitness.

Leg Lifts

This is one of the most effective exercises. They promote blood flow to your thighs and pelvic area and stimulate and strengthen these muscles.

For this exercise you will need to raise your body from the floor in an upright position. You could do this by holding on to a bar over your head, or better yet, propped on your elbows in a stand, such as the Nautilus leg exerciser often called the Roman Chair.

The object of this exercise is to raise your knees as close to your chest as you can, for up to thirty continuous repetitions. You will feel the effect of this exercise mostly in your thighs and pelvic area. With regular practice, you will strengthen these areas and improve your blood flow. Do this exercise at least three times a week.

The Kegel Exercise And Your "PC"

This exercise, in addition to improving your blood flow, has even helped some men gain control over premature ejaculation and, often results in a more powerful, more pleasurable orgasm. That alone, even if you have normal erections, may be reason enough to practice it! The exercise is named after Dr. Arnold Kegel, who first introduced it in 1948. Since then, thousands of men who practiced it, have benefited in a number of ways including bladder control, improved sexual pleasure and enhanced erectile function.

To perform this exercise, you need to locate your "PC"—pubococcygeus muscle. (Men usually exercise the "PC" only when it is contracted during ejaculation.) *Here's the method that is most likely to work* — pretend you are about to urinate or have a bowel movement and you are trying to hold it back until you can find a toilet.

You should feel a tightening around your anal opening and between your legs at the base of your penis. These are the muscles to work on. Regular exercise of these muscles strengthens them and improves blood flow to the penis, resulting in a firmer erection.

As you become good at locating and exercising your "PC" you can enhance the effectiveness of this exercise by doing the following.

While in a state of erection, locate the muscles that move your penis up-and-down and from side-to-side. Next, practice flexing these muscles and the "PC" at the same time. You will probably experience flexing of the thighs and buttock as well. Keep in mind that you are increasing the blood flow to this crucial area, which improves erectile functioning. So make it your intention to really give yourself a good workout!

The benefit of the Kegel exercise is improved muscle tone and blood flow. Men wishing to improve erectile functioning are encouraged to practice it on a regular basis.

You can practice these exercises almost anywhere, even while riding in a car. Incidentally, the Kegel exercise is beneficial to women as well, often resulting in stronger and multiple orgasms for them.

To summarize, the benefits of this exercise are:

- Improved blood flow to the penis and therefore *harder erections.*
- Control of *premature ejaculation.*
- More powerful and pleasurable *orgasms.*

Here's How To Do The Kegel Exercise

Begin by squeezing, as previously described. Hold this for about 10 seconds and then relax. Squeeze and relax. Start with your "PC" relaxed and then gradually squeeze as hard as you can, holding it for ten seconds. Do this for at least fifteen minutes or more, if you still feel comfortable. As you gain more control, you will find many opportunities to do the exercise, even when you're sitting.

Another variation is doing "quick snaps" of the "PC". These are done in the same way but quickly. Squeeze and let go to a rhythmical beat, like music or even your own heart beat.

Still another variation is the "squeeze and hold technique"–squeeze your "PC" as hard as you can and hold it for up to 30 seconds. Squeeze even harder just before you let go. Then relax for 15 seconds. Continue these repetitions for a total of 15 minutes.

Be sure to do the Kegel exercise in an organized manner. Use a chart (as is shown in Appendix "A" on "Performing the Kegel Exercise") to plot and organize your exercises, and plan on doing them twice a day for six-to-eight weeks. Thereafter, you may want to continue a less intense maintenance program. The potential benefits from a dedicated Kegel exercise program can be a tremendous aid in your ability to make love!

"For a long time now men have been searching for ways to increase their sexual vitality—in effect to increase their blood testosterone level, usually through eating or drinking something."

Chapter 4

WHAT ROLE DOES TESTOSTERONE PLAY IN YOUR SEX DRIVE?

"The search for something to increase testosterone level goes on, and new products are continually coming on the market."

How Does Testosterone Affect Libido?

Lack of sexual desire may be an indication of low levels of testosterone. Teenage boys and young men often seem to have an overabundance of this hormone. This may explain why some of them seem to be in an almost continuous state of sexual arousal. It's not uncommon for some members of this group to almost continually have a "hard-on." Certainly for most of them, getting an erection is no problem. Furthermore, it is possible for them to perform intercourse and be ready to perform again within a very short time—sometimes even within minutes! Such is the power of high levels of testosterone.

It has even been reported that pregnant women carrying male babies are often influenced by the testosterone of the child they're carrying. It gives them unexpectedly high sexual arousal. Women's testosterone need is only about 10% of that of men. With that amount, a woman's libido is active with no secondary male characteristics such as facial hair.

Testosterone is an essential ingredient in the proper functioning of the libido. Without it, the signal from the brain that produces an erection doesn't reach the penis.

What If Your Testosterone Level Is Too Low?

There is only one way to tell for sure and that is through a blood test, which is usually ordered by a medical doctor. When most men notice that the things that used to turn them on are causing much less of a response in their penis than in their brain, they'll conclude that they must need "something." If they are informed, they will suspect that the "something" they need is testosterone.

For a long time now men have been searching for ways to increase their sexual vitality—in effect to increase their blood testosterone level, usually through eating or drinking something. The search goes on even today and new products are continually being identified and coming on the market. Sometimes they are based on recent scientific research and work for some men. Most often, however, they are merely a variation of the usual combination of herbs, nutrients and plant extracts that men have been trying for years.

Damiana (Turnera Aphrodisiaca), Catuaba (Juniperis Brasiliensis), Sarsaparilla (Smilax Papyracea), Nettle Root (Urtica Dioica), Kola Nut, Muira Puama, Saw Palmetto, Ginseng, Extract of Green Oats, Lecithin, Prickly Ash Bark, Capsicum, some vitamins (usually Niacin, Folic Acid, B6, etc.), Zinc or Zinc Oxide, and L-arginine are usually on the list of things that work for most men. Just be sure of the purity and strength of the ingredients.

If your name ever gets on a mailing list for these kinds of products, you will receive a great number of solicitations. Most of them will make claims of almost miraculous reversals of impotence backed up by anecdotal proof and testimonials. Many will contain an unmistakable erotic element—predictably a cheap and tawdry one. What the solicitations most likely will not contain is a detailed list of the ingredients. In nearly all cases, the ingredients will turn out to be nutrients such as vitamins and minerals, along with herbs and other botanicals (as mentioned before, the "usual combination")—and these products are not without some merit in restoring male vitality and virility.

The obvious and most important concern about these products should be for the integrity of their manufacture. How can you be sure they contain the claimed ingredients? And, are the ingredients pure and in a condition that the body can assimilate and use? These are the questions that need to be asked about any vitamin or nutritional supplement. Promises of their curing impotence should be critically evaluated. The one thing that the products will not contain is testosterone itself—though I make this statement with some hesitation.

There actually was an over-the-counter product available in health food stores during the 1980's that contained a tiny amount of synthetic testosterone. Many men praised its effectiveness and, as you might expect, it gained a large and loyal following—until it was taken off the market. The product also contained a number of other ingredients that helped in achieving a stronger erection. It is reported to still be available in places other than the United States—Mexico, for example. (The product, called Zumba, is discussed in Chapter 3 and 4.) The many men who were self-treating with this product reported good results including stronger, harder erections, increased desire and more pleasurable orgasms.

So what's wrong with taking something that can yield such satisfying results? Most medical professionals will caution that taking testosterone is best done under the supervision of a medical doctor since side effects are possible. Nevertheless, you may wish to take other steps to assure that you maintain an optimum level of testosterone in your body. The potential gain in sexual vigor could be quite satisfying.

How Can You Enchance Your Erections By Improving Your Body's Natural Production of Testosterone?

Before you consult your physician about possible low testosterone, there are some things you can do on your own to boost your body's natural production of it. For example, if you are smoking or drinking excessive amounts of alcoholic beverages, stopping could result in significant benefits to not only your sexual health but to your health in general. *Your body's ability to produce testosterone is adversely affected by heavy or chronic use of nicotine and alcohol.* Take an inventory of your lifestyle habits. Determine whether or not you are doing things to yourself that can slow your body's production of the hormones necessary for sexual fulfillment.

Here are some questions to consider…

- Are you frequently having more than one alcoholic drink a day (one beer, one ounce of liquor or one glass of wine)?
- Do you smoke?
- Do you have a lot of stress in your life? Surprisingly stress reduces testosterone output.
- Do you suffer from depression? Some authorities report that depression is even more detrimental to testosterone production than stress.
- Do you live a sedentary life with little or no physical exercise?

If you answered "yes" to any of these questions, you need to recognize that you have identified a problem area or areas in your life that you can do something about. It requires taking responsibility for your situation and making changes in the way you live—"your lifestyle." The benefits from the changes I'm going to recommend, will also have a positive influence on your general health. Ideally, they will result in the optimum bodily condition for the healthy production of testosterone, which can, in turn, influence the quality of your erections!

Here are some things you can do to achieve the best physical condition for improving testosterone level and your erections…

What About Alcoholic Beverages?

If you drink alcoholic beverages, limit yourself to about one drink a day. Drinking too much has a number of negative effects on a man's lovemaking equipment. First of all, it adversely affects the delicate nerve network that's responsible for getting and maintaining an erection. Secondly, continual alcohol abuse will cause a significant drop in testosterone level and could, in the long-run, nearly emasculate a man. To be sure you aren't harming your ability to maintain healthy testosterone levels, *restrict your drinking to no more than one drink a day.*

While it is probably OK, on occasion, to have more than one drink, the best thing would be to quit entirely! Many moderate drinkers who have totally quit have been pleasantly surprised by the benefits they've experienced—which have included stronger erections and an overall feeling of well-being—and even, for some, better control of an overweight problem. However, modern day studies seem to suggest that one drink a day could be potentially beneficial to one's cardiovascular system. So, whether to cut down or stop all together is a decision each man will have to make for himself.

Among the things you can do to wean yourself away from chronic consumption of alcoholic beverages, one of the most effective is to *redefine yourself as a healthy person who is in charge of his own life.* Recognize that the feeling of intoxication from alcohol is an illusion that has hidden price tags. The "high" of alcohol is an illusion that you will awake from and be no better off and very possibly worse off than before. Certainly, if you achieve intoxication, your ability to perform sexually will be diminished. As Shakespeare had one of the characters in *Macbeth* say, "It provokes the act but kills the performance."

What About Smoking?

Medical research has found that smoking damages the blood vessels that a man needs to get and keep an erection. Smokers are four times more likely to be unable to get and keep an erection than nonsmokers just based on the damage that smoking does to the circulatory system and blood vessels. *The more you smoke, the more you are adversely affecting your sex life.*

Don't be fooled by the glamorous Hollywood-like portrayals of sexy smokers. Even cigar smokers who *seem* to be immune to the usual ravages of cigarette smoking are subject to the risks of tobacco as a carcinogenic agent. It's not unusual for someone to get away with breaking sound health rules for a while. But statistics do not favor smokers—there are few or no exceptions to the rules, at least not for long anyway. Those sexy smokers may seem to be OK, but obviously they are engaged in potentially self-destructive behavior and the odds are against them.

Recently, a staff writer at the *Los Angeles Times*, taking a cue from the California Department of Health Services, wrote, "What Viagra may give tobacco taketh away." The California Department of Health launched a campaign of television commercials on June 1, 1998 making just that point. You might call it the "limp cigarette ad." In it, an otherwise debonair handsome man, in the company of an elegant woman, lights a cigarette—which promptly goes limp. Smirking and shaking her head, the woman walks away. The message is clear, but the commercial includes a narration to emphasize the point—"Now that medical researchers believe cigarettes are a leading cause of impotence, you're going to be looking at smoking a little differently."[4]

The most direct effect of smoking on a man's sexual performance is that it constricts blood vessels, and thus can prevent a firm erection by restricting blood flow to the penis. And here is another important consideration. Although the evidence of the adverse effect of smoking on hormone production is less well-documented, some doctors believe that hormone production, including that of testosterone, is diminished by smoking!

So giving up smoking seems a rational thing to do no matter what your goals are. And there is little doubt that smoking is a potential block to optimum erectile function. The evidence proving the hazards of smoking is overwhelming. Choose to have a great love life that lasts; give up tobacco!

What About Stress?

Stress short-circuits your body's precise clock-like workings. Under stress, your body chemistry gets out of balance. Your immune system slows. People under stress are much more likely to fall ill. It's not surprising then that people under stress are less likely to enjoy sex fully. In fact, some people under stress may not be able to perform sexually at all.

Ultimately, *stress is really something you do to yourself!* I know, that for many, this idea is not easy to accept. You may recall situations that caused you stress in which it seemed that in no way could you have avoided feeling stressed no matter how much you may have wanted to. Nevertheless, the symptoms we call stress are all generated by the mind of the person stressed, and are therefore within the realm of his possible control. After all, *it's not what happens to us that matters, but how we respond to it!*

In situations of extreme danger or threat, no one expects to avoid fright and its accompanying stress. But, it is well within the capability of most people to manage their *reaction* to the ordinary day-to-day situations that produce stress. It is beyond the scope of this book to deal with stress management, but if you have a need for it, there are many sources of information available. For now, remember that the means for dealing with the day-to-day stress that accumulates and interferes with maintaining a vibrant and healthy body, and an optimum testosterone level, is reasonably within most men's capability. If, however, you find that you are not able to deal with your stress on your own, then the wisest thing you can do is to consult a mental health professional or enroll in a self-help program of stress management.

What About Depression?

Depression robs your body of vitality. Like stress, it interferes with your body chemistry and slows down your immune system, only more so! Depression, in most cases, is also basically something you do to yourself. To the depressed person, however, this may not seem to be so. However, there are cases in which physical and mental conditions in the body can result in depression. But typically, depression results from a person's mental reaction to a situation they find unacceptable and is, in that sense, self-induced.

You may feel that you can't help feeling depressed, and that you are not responsible for it. But actually, the more you accept this idea and consider yourself a victim of depression, the more susceptible you become to suffering from it. If you're depressed, you can't reasonably expect to have a scintillating sex life. Even mild depression interferes with a man's testosterone level.

The good news is that depression, like stress, as many experts agree, is a problem you can do something about. It is entirely possible to manage your mental state so as to simply reject depression outright from the start. Although some people can do this, depression for others can be a seriously debilitating affliction which they cannot cope with on their own. In that case, it is imperative that a qualified psychologist or psychiatrist be consulted. Nevertheless, the point I want you to recognize is that *depression, even mild depression, interferes with a man's ability to have a firm erection.*

It's easy enough to understand that if you are depressed you're likely to have a decreased desire to be sexually aroused. This may have little to do with the fact that depression is limiting your body's ability to produce that important hormone, testosterone. Still, it's not difficult to imagine that a man suffering with depression, particularly mild depression, may desire to have sex—and indeed many do. Unfortunately, failure to get and keep an adequate erection is more likely to occur with such men. But worse still, such a failure is likely to reinforce the depression, resulting in an even more limited production of testosterone. Simply put, *depression works against having a firm erection by slowing the body's production of testosterone.*

Again, it is beyond the scope of this book to deal extensively with managing depression. However, there are also many excellent sources of information available on this subject as well. Any man who recognizes that depression is limiting his sex life should take the initial step of getting informed. Remember that, in many cases, it is possible to learn to manage depression on your own. And, if you cannot manage it on your own, you must honestly admit it and seek the help of a psychologist or psychiatrist. By whatever means you overcome depression, you will greatly improve your life and your erections—the gratitude of your lady will be enormous!

What About Exercise?

Just as exercise is the key in improving blood flow, as we discussed in Chapter 3, it is also a key to maintaining an optimum level of testosterone. It may come as a surprise to you that vigorous exercise actually improves the body's production of testosterone. The recommended level of exercise needed to maintain an optimum fitness level, as discussed earlier, is to induce the "training effect" in your body for fifteen to thirty minutes, five times a week. This translates into something equal to about fifteen to thirty minutes o` calisthenics, five times a week. *You can improve your love life significantly when you get into a vigorou: fitness program.* But first check with your physician, especially if you've been sedentary.

Don't be discouraged if you are adamantly opposed to vigorous exercise. Even sedentary people can make very satisfactory improvement, as was also previously stated, by simply taking up walking! An hour of walking, three or four times week, will do nicely!

But I must again emphasize that it is a vigorous exercise program, including both aerobic and muscle-development exercises, which will result in the greatest improvement of your body's production of testosterone. Aerobic exercises are those that cause a sustained, significant increase the heart rate, such as calisthenics and running. Muscle-development exercises include weightlifting and using resistance-type exercise machines like those manufactured by Nautilus. Two things happen that benefits attaining better erections with these two types of exercise. You attain better blood flow and you stimulate the body's production of testosterone.

The muscle-building exercises are necessary because they stimulate the production of the receptors needed to make testosterone useful. So you need both aerobic and muscle-building exercises for the twofold reason that you achieve better blood flow, and you increase the body's production of and its ability to use that all-important male hormone, testosterone.

A man who jogs or participates in a sport like racquetball or tennis, and who also lifts weights would be filling the prescription for better blood flow and testosterone production nicely. The idea is to participate in activities that *raise the pulse rate and sustain it for at least 20 minutes* (attaining the "training effect") and also to *do exercises that strengthen the muscles.*

You will recall similar advice from our previous discussion on exercise. The program that will benefit the production of testosterone must include sustained pulse-raising exercise and muscle-development exercises, such as weightlifting.

What About Testosterone And Dieting?

There's at least one other hazard to the body's production of testosterone – "dieting." Medical doctors have confirmed what many athletes have long-suspected —rapid loss of excessive amounts of body fat can cause a temporary loss of sexual prowess. While testosterone is normally being made and broken down in the body on a continuing basis, under conditions of severe restriction of available fat in the body, like going on a starvation diet, the body shuts down production of testosterone. A certain amount of fat is needed for the production of testosterone, and that fat is simply not available when the body is reacting to a starvation diet.

While the importance of weight control to good health and a vibrant sex life is obvious, the body still requires a small amount of fat. The problem starts if one goes on a very restrictive or near-starvation diet. The body responds as if it were actually in danger of starvation. It shuts down or curtails production of a number of hormones that need fat—among them testosterone. Actually strenuous overtraining can have the same results—as experienced by some athletes, e.g., long-distance runners. Under normal conditions, even thin men will have enough fat in their bodies for the optimum production of testosterone, but starvation dieting and overtraining are not normal conditions.

Are There Things You Can Eat That Will Improve Your Testosterone Level?

There are a number of natural substances which have been used by many to stimulate the body's natural production of testosterone. While it cannot be definitively stated that all of them increase testosterone levels, the testimonial evidence that they've helped men overcome sexual deficiencies is very compelling. At the very minimum, they have been a sexual tonic for some men, either alone, or in the case of vitamins C and E for example, in conjunction with taking other nutrients and tonics.

This list includes the following…

- Damiana
- Saw Palmetto
- Extract of Oats
- Raw Pumpkin Seeds
- Ginseng
- Vitamin E
- Lecithin
- Vitamin C
- Withania Somnifera
- DHEA (delhydroepiandosterone)

Damiana, the dried leaves of a Mexican plant, which are easily obtained, are legendary as a sexual tonic. They are usually boiled and made into a tea; however, capsules are available in many health food stores as are the leaves themselves. The benefits are said to include a general improvement in vitality and a sense of well-being. This would be the expected outcome of an increased testosterone level.

Saw Palmetto *(serenoa repens)* is a small palm tree found in the West Indies and along the Atlantic Coast of North America. This tree produces dark reddish-brown berries from which fruity smelling oil can be extracted—beneficial to male sexual health. The dried berries themselves can also be purchased. These berries or the extract from them have been prescribed for low libido, impotence and even atrophy of the testicles. It is likely that the good results from this botanical is because of its influence on the increased production of testosterone.

Saw palmetto has a well-deserved reputation as an herbal cure to shrink enlarged prostates. (An enlarged prostate is a medical malady known as BPH or *benign prostatic hypertrophy*.) The connection of the prostate gland to testosterone is quite direct. It's a mutated form of that hormone, dihydrotestosterone (DHT), that causes the enlargement. Scientists believe that Saw Palmetto deactivates an enzyme (called reductace) which causes testosterone to mutate to the harmful form of DHT. It also has a healing effect on the prostate gland.

Acting on this knowledge, a reported 2000 BPH patients in a study in Germany were successfully treated with Saw Palmetto according to *Prevention Magazine* herbal advisor Varro Tyler, Ph.D. Extracts of the plant are widely available in European pharmacies, specifically to reduce enlarged prostates. The same or similar products can be found in health food stores in the United States.

Extract of Oats generally refers to a European product that consists of an extract of green oats—reputed to increase sexual desire. "Feeling your oats!" is an old expression that has been scientifically shown to be valid. Reports of its effectiveness on a television magazine show caused quite a stir. A formulation from a manufacturer in Switzerland is available in capsule form in many health food stores. My own opinion is that unrefined oat cereal may be almost as helpful without the high cost. At any rate, studies indicate that these oat-based products stimulate increased sexual function in some people, especially when combined with extract of nettle and some other botanicals.

Raw Pumpkin Seeds *(pepitas)* are sometimes called a male superfood! They are a good source of quality protein and zinc, magnesium, and essential fatty acids, which, in themselves, are beneficial to the body's production of testosterone. Over and above that, they are believed to contain vital trace elements that stimulate testosterone production.

Ginseng, the root of a Chinese herb, which is also cultivated and highly acclaimed in Korea, has a worldwide reputation as an energy enhancer, of particular benefit to males. Russian athletes were reported to be using it on a regular basis during the heyday of Soviet success in the Olympics. Two types or varieties are usually available, Siberian and Korean, but buy ginseng carefully! Get assurance that you are receiving a product that has the beneficial elements needed for the much heralded tonic-effect of this product.

The strength and purity of ginseng available in health food stores and even in Oriental markets varies greatly. Not surprisingly price is sometimes an indicator of quality. The older ginseng roots sell for higher prices and likewise contain more of the beneficial elements.

Many people have reported increased libido and better sexual performance after taking ginseng. However, ginseng, as a sexual tonic, is reputed to be more helpful to older men than to younger ones. This perhaps indicates that ginseng influences testosterone level in the body or influences the availability of testosterone in a form that will improve sexuality. Younger men, unlike older ones, are less likely to be deficient in this regard. Ginseng is available in a number of forms, including extracts, capsules and the root itself. The root is usually available dried and it is used to prepare a tea by boiling in water.

Vitamin E is long recognized for its role in reproductive health. It is also an essential nutrient in male sexual health. It's benefit as an antioxidant is much publicized.

Pycnogenol—This antioxidant patented formulation of the extract from a French pine tree is thought to have a positive effect on the chemistry involved in attaining an erection. If, as speculated, it has a positive influence on the presence of nitric oxide in the penis, then it should increase the effectiveness of Viagra. If you should take it in the hope that it does, you can nevertheless be assured that it will be beneficial as an antioxidant much the same as vitamin E, C, and A.

Lecithin is a nutrient found in many foods, but manufacturers often include it in the formulation of supplement products intended to improve sexual function. The reason they do, no doubt, is because it nourishes nerve and cerebral cells.

Therefore it may improve delivery of that important sexual desire signal from the brain to the nerve cells of the penis.

Eldeprnyl, a drug, (sometimes called L-deprenyl) is said to be almost too good to be true, but it is difficult to obtain. It's called a "sexual energizer." It is prescribed mainly for patients with Parkinson's disease, with its effect on sexuality considered a side effect. There is speculation that, in low doses, it could be used to enhance and increase sexual desire and performance. Moreover, don't expect to be able to obtain it.

Vitamin C —Yes, this venerable old standby also has a reputation as an aphrodisiac. Certainly this vitamin, which plays such a vital role in maintaining the health of the immune system, is beneficial to erectile functioning. Like vitamin E, it is recognized as an antioxidant. Moreover, it is likely that nearly everyone recognizes this nutrient as an essential and necessary part of a healthy daily diet.

Withania Somnifera, an Indian herb, is reputed to be excellent for treatment of sexual deficiency in men. It has only recently been available in health food stores in the United States. It is fair to say that its beneficial effect involves an influence on testosterone level or its availability. Be advised that its Indian name, *ashwaganda,* means "smells-like-a-horse!"

DHEA (delhydroepiandosterone) is being touted by many and taken by millions, both men *and* women. This hormone, like testosterone, is produced naturally in the body. DHEA in a synthesized version is now widely available in pill form, unlike testosterone, without a prescription. Its effect is something like that of an aphrodisiac. There is little doubt that if you are deficient in this hormone, it will have a pronounced effect, not only on your sexual functioning but, on your general feeling of well-being. It may also increase you libido. But taking it on your own is pretty close to taking testosterone itself without the supervision of a doctor, which is not recommended. The manufacturers seem to be implying that it is safe.

Women especially are reporting good results with DHEA because it is reported to increase desire, particularly in menopausal women. Note that, *most doctors are advising against it for men at risk for prostate cancer.* Generally speaking, that might include any man over the age of 50. More specifically, it would include older African-American men. Statistics show that they have the highest incidence of prostate cancer of any ethnic group. *Men in which a family member has had prostate cancer, and definitely men who have been diagnosed as having prostate cancer, should stay away from DHEA.*

Many health professionals are urging caution on taking DHEA because of the risk of creating a hormonal imbalance in the body. Anyone taking supplements of a hormone that the body produces naturally, runs the risk of interfering with the body's natural production—by slowing it down. While DHEA could have a very pleasant positive effect on your sexual and overall vitality, it would be much wiser to take nutrients that, instead, enhance or nourish your body's natural production of these hormones.

All of these products can usually be found in health food stores in formulations intended to enhance male sexual function. They are often recommended both for increased libido and stronger erections.

By carefully applying this information, you could stimulate your production of testosterone and shift it into high gear. Your sexual sense of yourself will improve, your desire will increase, and you will find sexual arousal more easily initiated and more rapidly flooding through your body (see Chapter 4).

But What If You Have An Abnormally Low Level Of Testosterone?

If you suspect that you may have an abnormally low testosterone level, you should see your doctor. A major symptom of an abnormally low level of testosterone would be no sexual desire over an extended period of time. Your doctor can make a determination of your testosterone level through a blood test. If it is low, he may prescribe testosterone in either an oral dosage or by injection directly into the bloodstream. Injection has been the preferred way to administer the hormone.

The oral dosage has been associated with adverse effects on the liver, in some men.

But advances are being made in the administration of the hormone. There is now available what many believe to be an even better treatment: *a transdermal testosterone patch*—a device similar to the nicotine patch. With this patch placed on your skin, as directed by your doctor, the medication is absorbed into your body. A California pharmaceutical company, Alza, has developed such a skin patch for treating testosterone deficiency in men.

With the introduction of the necessary testosterone into your body—whether by injection, oral dosage, or patch—your sexual proficiency should dramatically improve. While testosterone is necessary, and even essential, for proper libido functioning in both men and women, it plays an even greater role in men in the expression of maleness and masculinity. Just the same, too much of a good thing, is *not* a good thing. Studies have shown that men with unusually high levels of testosterone tend to be overly aggressive. They alienate others, have higher divorce rates, and tend to be lonely and dissatisfied. Athletes who take supplemental testosterone for rapid muscle development and increased strength run the risk of dire health problems. It is for this reason that steroids are condemned and almost universally banned by athletic organizations!

"Ya gotta do what ya gotta do."
Sylvester Stallone

Chapter 5

ENHANCE YOUR ERECTIONS THROUGH NUTRITION AND MEDICINE

"Many people agree about the beneficial effects of these products."

A Variety Of Choices Are On The "Menu."

It is amazingly important to every man that his penis gets hard when he thinks it should and that it functions the way he thinks it is supposed to. This is a concern of men, from the time of early sexual awareness in their teens, through their senior years. It's no wonder then that there's a great store of lore on things to eat, drink and swallow to assure or enhance sexual performance and enjoyment. They include everything from glandular extracts (e.g., testicle, prostrate, pituitary, and adrenal) of domestic cattle and also other animals, to an extensive array of herbs and other plants. There are also some exotic and expensive substances, like powdered rhinoceros tusks, some of which are illegal. (Rhinoceros tusks are definitely illegal!) Interestingly, some of the available products work to some extent for some men. Some work quite well and you may want to take note of these.

Oysters are legendary. While their effect may not be immediate, eating them is worthwhile to your sexual vitality.

They are an excellent source of zinc, a necessary nutrient that promotes sexual health, which influences a man's ability to achieve an erection.

Here is a list of some products. Some are less well-known, and some we've already discussed in the previous chapter, concerning their effect on the body's testosterone level and their other reported benefits as well as the cautions. There is a lot of agreement about the beneficial effects of these products: **Ginseng, Saw Palmetto, Extract of Oats, Damiana, Raw Pumpkin Seeds, Vitamin E, Lecithin, and DHEA** which were covered in Chapter 3.

Extract of Nettle —Usually an extract of the root of the Stinging Nettle is beneficial to male sexual health.

Fava Beans —Naturally contains the substance **L-dopa,** known to increase sexual drive and desire when given in prescription doses (mainly prescribed for patients with Parkinson's disease). It increases a neurotransmitter, dopamine in the brain, which is intimately linked with sexual drive. Dopamine is also associated with facilitating memory, energy, a sense of well-being. The drug is available only by prescription, but Fava Beans can be found in the grocery store. A sixteen ounce can is said to contain as much as a prescription dose of L-dopa.

Sarsaparilla, as a tea, has had a long-standing following as a sexual arouser or tonic. It is interesting to note that in the past it has been dismissed about as much as it has been accepted. Yet today, it is formulated as *smilax officinalis* and is used by body builders to increase energy and endurance. Yes, this same root, which is used to flavor teas and soft drinks, is recognized for its beneficial effect on the hormonal system.

Zinc has to be accepted on the face of it as a beneficial nutrient to male sexual health. It is, after all, the most critical nutrient needed by the prostate, which needs ten times more zinc than any other organ in the body! It is best taken in the form of zinc picolinate.

Gingko Biloba, which is being taken today by millions to increase mental alertness, also stimulates circulation to the extremities. Any improvement in circulation must be counted as beneficial to erectile functioning since an erection is precisely the result of increased blood flow to the penis. Men using gingko biloba report an improvement in their sexual functioning.

Muira Puama has made an appearance lately as a product that will benefit male sexual functioning. Many claim that it has a dynamic effect on sexual desire and performance with no side effect. Muira Puama has been touted as a very effective traditional Brazilian aphrodisiac, but is believed to be beneficial in treating a number of other ailments including rheumatism, stress, frigidity, menstrual cramps and premenstrual syndrome (PMS).

If claims for this botanical are to be believed, it increases the body's natural production of testosterone. Users, both men and women, confirm that the claims for it as a general tonic are valid in that, while using it, they experience improved strength and stamina. As one female user puts it, "It really makes me feel frisky!" A middle-aged male user reported that he had a "wet dream" the night after taking it that morning. "And, it was the first in a long time," he reported.

Best of all, there are no reported side effects. It is usually available in health food stores as an extract and sold in a bottle with a dropper. The directions generally advise taking a few drops in a small amount of water.

Yohimbe—*this may be the only product recognized by the medical profession as an aphrodisiac*. This substance originally came from a tree in Africa. It is now a readily available pharmaceutical under the name *yohimbine hydrochloride*, and it is also available as a natural product in health food stores. It improves erections by its influence on the flow of blood into and out of the penis. For most men it works quite well and it is being prescribed by doctors for exactly that reason. A medical doctor's prescription may also include *Trazodone* which, when taken with yohimbine HCL, enhances its effectiveness. The word is out on yohimbe and, even now, thousands of men are experimenting with it. It has few side effects which only affect some men, and even then, they appear to be transitory.

What About Formulations of Herbs and Traditional Medicines?

Lately there have been strong claims about the effectiveness of a formulation of Chinese herbs and traditional medicines. One such product, *Strenixx*, available by mail or telephone, claims to be "an original, all-natural male potency formula." It is said to be a blend of at least nine herbs, presumably of Chinese origin, and according to its publicist, LTA Media, it is manufactured by a company called Pharma Botanical. Presumably, also it is based on centuries-old Chinese experience with herbs. If this is true, it could have positive implications for improvement of ED. However, to my knowledge, no clinical trials have been reported. Anecdotal evidence, however, does seem to give strong support to this product.

I mention this product to alert you to the ever- increasing introduction of products of Asian origin appearing on the market worldwide. No doubt many of them are based on a solid tradition—particularly in the case of Chinese medicine. But careful scrutiny, and even professional advice, may be required in evaluating such products.

Except for the Eldeprnyl and L-dopa most of these products are easily available. Some can be found in health food stores in formulations especially intended to enhance male sexual function. They are often recommended both for increased libido and stronger erections.

One product that gained a great deal of notoriety before it was taken off the market in most or all U.S. cities was a formulation of damiana, lecithin, and, most importantly, a synthetic testosterone and yohimbe. A lot of men considered this product "dynamite"—the real thing! Its formulation suggests that it should have had an unmistakable positive effect and for many it did. Yohimbe, one of its main ingredients, has gained wide acceptance as beneficial to men who want better erections, and its tiny amount of synthetic testosterone (its other main ingredient) could have a stimulating effect on the libido.

But selling synthetic testosterone over-the-counter, even in the tiny doses contained in this formulation, raises a lot of questions. The product was sold under the brand name, "Zumba." I have been informed that it is still available over-the-counter and by mail order from pharmacies in Mexico. *Although testosterone is sometimes prescribed by medical doctors, the best advice is not to take it unless you actually get a prescription from your medical doctor* (see Chapter 4 on testosterone).

Some Interesting Case Histories

But here are some interesting case histories on the use of this combination of substances from Dr. Bolton's files. Note that all names used in these descriptions in this book are fictitious so that confidentiality is maintained. Also, some of the circumstantial facts have been altered somewhat, which means that if the descriptions match people you know, it's purely coincidental.

Ed, a thirty-nine year old white-collar manager, was concerned that he did not always achieve a firm, satisfactory erection. He believed his problem was psychological and he was seeing Dr. Bolton for counseling when he, on his own, discovered this product which he called Zumba. He reported that he got stronger erections that lasted longer. Later it was determined that Ed's problem was stress and when he got his life together so as to eliminate or control it, his erection problem vanished! Still there is no reason to doubt that this product helped. Fortunately, he learned to manage the stress in his life because this product soon became very difficult to obtain.

Mark, another of Dr. Bolton's patients, also experimented with Zumba. Forty-four years old, he used the product occasionally for a week or two at a time. He was troubled because his erections were only what he called "half hard" and sometimes he could not get an erection at all. According to Mark, sometimes the product worked and sometimes it didn't. He later consulted a urologist and found the reliability he wanted through vacuum therapy.

Both Ed and Mark had, through happenstance, stumbled upon one of the most remarkable male aphrodisiacs available today, the yohimbe contained in the Zumba. Furthermore, their experience illustrates that a variety of approaches can be helpful in assuring that you have a ready erection when you want one. Finding the right substance to eat or drink to enhance erections can be challenging since each man's nutritional or therapeutic needs are different. There's one substance, the one that Ed and Mark found, yohimbe, that is reliable for most men. In fact, it is recognized by the medical profession as an effective product for improving erections.

Yohimbe, as discussed above, was included in the formulation of Zumba, so it's obvious that there is wide agreement that it is effective. It is available by prescription and also over-the-counter at health food stores. Yohimbe, as you'll recall, is obtained from the bark of a tree said to originate in Africa. At present, it is also obtained from related trees in various regions. *Many men have had their erection problem alleviated by a doctor's prescription for yohimbe.* Others, who have simply purchased over-the-counter yohimbe at a health food store, have also had strikingly good results. But be wary: the strength and purity of over-the-counter yohimbe from health food stores may vary not only from brand-to-brand but even in different batches of the same brand. Check with your health food store merchant for the name of the manufacturer with the highest standards of quality control.

Quality control is a crucial issue in evaluating alternative health treatments. Even if a product has the capacity to be effective as a treatment or therapy, its effectiveness will be influenced by its purity, strength, any adulteration, and other ingredients it may be compounded with. This issue is being addressed by some manufacturers of "health food store" products. For example, a Southern California biotech company, Pharma-Print, has developed a method to identify the active component of herbs and medicinal plants. They are clinically testing herbal remedies to verify their effectiveness. This could result in standardizing dosages. But more than that, it could result in the development of more effective botanical and other natural products and the weeding-out of ineffective ones. When that day comes, we will not have to rely solely on the testimonials of others. Pharma-Print is reportedly, close to introducing a standardized extract of the saw palmetto berry, which is used for treating prostate problems, as discussed earlier. If Pharma-Print achieves its goal, it will develop pharmaceutical quality versions of herbal and other botanical remedies.

What About Bob And His Quest For Stamina And Readiness?

Finding things to eat or drink that improve a man's ability to have an erection requires careful self-observation. If the concern is about occasional erectile failure, it may be hard to know whether it was the good night's rest or the pill you took that did the trick. Still there are many men who follow a regular regime of taking nutritional supplements and they're convinced that they make a difference. A very good example from the files of Dr. Bolton is that of Bob . It was important to him that he always be at his peak of stamina and readiness. It is not surprising then that he had an array of supplements. They included glandular extracts, oysters, ginseng and vitamin E. It worked—both he and his wife were thrilled at their love-making success.

Bob gave the most credit to "glandulars." Glandulars are usually available as pills that contain tissue from glands of food animals. Experience seems to demonstrate that some factors beneficial to humans are available in these tissues when properly prepared in modern laboratory conditions. For example, *orchic* tissue from the testicles, manufactured into pill form, improves the sexual vitality of males as an aphrodisiac. But those who advocate taking glandulars also include several other tissue sources, particularly those of the endocrine glands, especially the adrenal and pituitary. The list of beneficial tissue sources include several other organs, some of which are not directly associated with sexual function e.g., the liver, heart, and pancreas.

If you seek out glandulars in your local health food store,

you'll probably find a formulation especially for males just as Bob did. It will contain tissue from a number of sources in order to provide the factors necessary for increasing sexual vitality. Some manufacturers also include nutrients such as selected vitamins and minerals. Glandulars may work for you as they did for Bob. Because they are basically benign there is little risk in trying them. The cost of a trial period of therapy will be modest. Most men who take glandulars for sexual enhancement start with or include *raw orchic* (testes). This seems to be a reasonable way to assure intake of the nutritional factors that may be associated with the production of testosterone. The raw orchic is possibly most effective when taken with a combination of L-Arginine, which is usually available in health food stores as amino acid or a protein supplement. Also, to be most effective, tablets of raw orchic should be the type that releases only in the small intestine and *not* the stomach.

Energy Enhancers For Improved Performance

There is another class of products that seem to improve sexual performance and arousal, including stronger erections. They are marketed as energy enhancers under brand names like "Rocket Fuel"! Formulations vary but generally they will contain substances from herbal and other natural sources. Typical ingredients include *gotu kola, ginseng, juniper berry, sarsaparilla.* These kinds of ingredients have been shown to have a tonic-like effect on many people, energizing them and improving their sense of general well-being. The reputation of some of them as a sexual tonic for men is legendary—this is certainly true of ginseng.

Although I have not seen any direct claim by a manufacturer that these products were meant to have an aphrodisiac application, some makers strongly hint that they do, and some men report that they have better erections as a result of taking them. They are most often available as a liquid in a vial with a dropper, but capsules and pills are available from some manufacturers. Although they enhance the body's vitality in the manner of a nutrient, they are essentially total body stimulants and work mainly through the nervous system similar to caffeine. They may also contain *cayenne,* which may possibly act as a mild urinary tract stimulant. I have personally tried "Rocket Fuel." It is easy to take—a few drops in a glass of water or fruit juice—and, for me, it works.

Saw Palmetto—Promotes Prostate Health And Better Erections, Too!

A product that has considerable support, even by some physicians is *saw palmetto,* as mentioned earlier. The extract of these berries is prescribed as a treatment for prostate ailments, especially enlarged prostate, as previously discussed. In this condition the prostate is enlarged, but the cause is not cancer. The symptoms are related to difficulty with urination. A man with this condition will find that relieving himself is slow, sometimes painful and sometimes incomplete. Urgent calls to "go" come with increasing frequency and little warning. And frequent trips to the bathroom at night interfere with getting a good night's sleep.

Research in Europe has established that this condition is relieved by standardized extracts of the saw palmetto plant available in pharmacies there. A typical dose is 160 milligrams in pill form, twice-a-day. It works and has negligible side effects. If you decide to take saw palmetto extract for prostate health or to improve your erections, you should, of course, be carefully observant for possible side effects. In any event, any man with symptoms that might indicate prostate problems should absolutely see a doctor to determine the cause. *Keep in mind that prostate cancer, if identified early enough, is curable!*

Saw palmetto's use in promoting prostate health is established, and recognized in the medical profession—particularly in Europe and especially in Germany. It is therefore reasonable to believe that improving prostate health will benefit sexual functioning since nerve connections in the prostate have a direct influence on the ability to achieve an erection. Furthermore, enlargement and inflammation of the prostate are conditions sometimes associated with low testosterone levels.

Evidence indicates that saw palmetto stimulates increased production of testosterone while inhibiting its breakdown into harmful forms of testosterone that cause enlargement of the prostate gland. Therefore, it has a healing effect on the prostate gland—in particular in cases of BPH (benign prostatic hypertrophy). In some countries in Europe, it is the first line of defense against this malady. Many also consider it to be a sexual tonic for men—but this is probably a "side effect" of its medicinal effect on the prostate gland. Even men of somewhat advanced years who have used saw palmetto have, in some cases, reported the return of their erections!

If you want to use saw palmetto, you need to decide what form you will use. The practice of doctors and patients in Europe indicates a preference for the pill version. But a liquid extract, usually sold in a small vial with a dropper, as mentioned before, is also available, which is equally as potent as the pill.

Because the berries of the saw palmetto shrub are so abundant, easy to obtain and inexpensive, some have tried to extract its benefit by brewing a tea from the berries. This won't work because the beneficial ingredient in these berries is not water-soluble. I would even advise against ground-up powder of the berries sold in capsules. The most effective way to attain the benefit of the active ingredient in these dried-up little berries is to obtain an extract. As always in the case of over-the-counter health food store products, be prudent. *Check to assure that it is manufactured by a reliable company whose statements about the content and purity of the extracted substance can be depended upon to be accurate and true.*

New ideas and approaches are continually being pursued both in alternative medicine and by established pharmaceutical companies. It is no exaggeration to say that the quest to develop a product that can be taken orally to produce an immediate erection is at peak momentum. Even at the time of the final preparation of this publication, a report in the *Journal of Urology* tells of such a substance which is to be absorbed through mucous membranes of the mouth. Reportedly, it is being tested for possible FDA approval. This medication or substance taken by mouth is said to have resulted in satisfactory erections, in a matter of minutes, for the men in the small test group. Incidentally, the substance in this pill is related to the pharmaceuticals used in injection therapy which we will discuss in a later chapter.

How About Harnessing Body Chemistry To Improve Erections?

Another idea receiving a lot of attention by commercial purveyors of "cures for impotence" is that a lack of a particular substance in the body is the cause of erectile disorders. This is a plausible notion actually supported by recent medical journal announcements that just such a substance has been identified. In this case the substance is *nitric oxide*. It is theoretically possible that a substance that is normally naturally present in the body could play a crucial role in erectile functioning such that, in its absence, an erection could not occur. It follows, then, that if that substance is not present or not present where it is supposed to be, having an erection is unlikely or impossible. The introduction of Viagra has established that the claims about nitric oxide were correct.

As it turns out, scientists have discovered and announced just such a substance—a complex chain of proteins that, when present, causes the relaxation of the blood vessels in the penis that produce an erection. The key factor in this is now known as nitric oxide. Scientists have only recently begun to understand the role of this mysterious substance as a vital element in several body functions. It only started to receive scrutiny in 1987. It has been referred to by some scientists as a "super molecule" because of the variety of effects it has on the human body.

And yes, as a result of some mysterious results of a neurotransmitter, the nitric oxide appears briefly in the penis. It causes the smooth muscle vascular tissue to relax under sexual arousal, thereby producing an erection. Researchers from the University of California at Los Angeles to the Johns Hopkins University in Baltimore, to the Wellcome Research Laboratories in England, are forecasting its use in treating a variety of conditions. These include a diverse spectrum as various as IBS (irritable bowel syndrome) and high blood pressure—and, of most interest to us, ED.

Nitric oxide appears to be a substance ordinarily thought of as a corrosive. Even so, there are actually products on the market, sold mostly by mail order, that claim to treat male deficiencies of it as a cure for "impotence." Certainly no one would actually ingest nitric oxide, because, of course if it's used improperly, it is toxic! Instead, manufacturers offer other ingestable substances that hopefully will foster the natural production of nitric oxide. It could then make its brief appearance in the penis, under sexual arousal, to play its role in the relaxation of the smooth muscle vascular tissue, causing an erection.

The companies selling these products present testimonials from users who proclaim its effectiveness. Some manufacturers have even presented an alternative to ingesting the lacking supplements. They opted, instead, to supply the missing substance to the outside of the penis in the form of a cream or salve to be absorbed through the penis. Again the manufacturers of these formulations are able to present testimonials from users who proclaim that the substance is effective.

The idea of presenting a substance to the outside of the penis to relax the blood vessels, and thus produce an erection, has its merits. It is even now being tested by medical doctors, as you will learn further on in this book. Keep in mind that the substances now being prescribed by medical doctors to apply directly to the penis must be either injected with a needle and syringe or deposited into the urinary tract. The idea that rubbing a cream or salve on the outside of the penis can produce an erection is plausible, and this is actually being tested by scientists.

There actually are things you can ingest that will have a beneficial effect on erectile function. Many, if not most, of them have only the testimonials of the men who have used them to justify the claims made for them. Some, however, have stood up under the scrutiny of scientific observation, yohimbe, for example. While others are basic nutrients that benefit body processes that have a direct influence on erectile functioning and sexual vitality. But read on.

Major pharmaceutical companies are now testing and introducing new pills that cure ED in a lot of men. New discoveries are continually being made. I have no doubt that the perfect combination of nutrients, herbs and other substances will be available for optimal erectile functioning. They will be discovered, just as science and medicine are continually finding new ways to successfully and directly treat impotence so we can truly achieve "potency-on-demand."

"It makes sense that the more open and honest your marriage relationship is, the more likely you are to share your feelings about your ED. You'll ask for your wife's support and understanding of the solution you've found, whether it be an injection or something else."

Chapter 6

INJECTION THERAPY

"One man discovered that, with this method he satisfied his wife's sexual needs more than he ever could have imagined."

You Can Have An Erection Anytime You Want To Even If Viagra Doesn't Work For You!

Here's a method that can give a man an immediate hard erection. No excuses, "if's," "and's" or "buts" about it. It produces results even if Viagra isn't for you! And the result is immediate and very satisfying. It puts the ability to have very fulfilling intercourse, without fail, into the hands of most men whose libido is at least minimally intact. "If you want to do *it*, you can."

It is not surprising that men who use injection therapy are consistently rated higher by their lady! There is the anecdote going around about the guy using this method who happened to get a peak into his new wife's diary. He discovered, to his amazement, that he satisfied her sexual needs more than he ever could have imagined!

Charles' experience, discussed later in this chapter, confirms the effectiveness of the substance used in this method. He was given a prescription for this pharmaceutical after his urologist diagnosed him as having intermittent or recurring mild impotence. Charles was astonished when the doctor injected him right there in his office and, *as if on command, his penis lengthened out into a stand-up erection.* During this visit to the doctor's office, Charles was given advice and information about the medication. But, before he was given a prescription, he was required to make a second visit for the purpose of demonstrating to his physician that he could inject himself.

Following his doctor's instructions, he actually injected himself. He reported, in his own words, that "It was a piece of cake!" Again, right before his eyes and those of his doctor, his penis rose up to a firm, hard erection. Charles wanted a prescription right then and there. His doctor obliged after a few words of advice and caution, such as about the frequency with which he should use the medication and how much to inject.

Reportedly, the pharmaceutical most often prescribed for injection therapy is prostaglandin. Early in the application of injection therapy, however, papaverine hydrochloride was used. In fact, papaverine is still sometimes used in combination with prostaglandin. Exactly which of these substances to use is a decision that you and your doctor would make. It's important that you talk with your doctor about the various substances that are combined in these kinds of prescriptions. Currently, various combinations and trade names are available at pharmacies. They include, as stated before, prostaglandin and papaverine but also phentolamine. Different strengths and combinations may be prescribed by your physician, but prostaglandin (or prostin) seems to be the substance of choice.

What If Viagra Doesn't Work For Me And I Hate Needles?

There actually are alternatives to injections using these or similar medications—for example—*urethral gel and oral phentolamine*. And many patients have successfully used them. The urethral gel is either inserted into the urethra through a special syringe or it's applied topically. Oral phentolamine is a medication which can be taken orally either by swallowing or placed under the tongue.[5] It may one day rival Viagra.

Getting Back To Prostaglandin—What Is It?

The name prostaglandin would seem to indicate that it has something to do with the prostate gland. In fact, "prostaglandins" were first discovered in semen which, of course, is produced by the prostate gland—hence the name.

It has now been established that prostaglandins occur in many tissues and are not restricted to any specific organs. Their effect in the body is similar to that of hormones, but different in that they stimulate cells near the locality where they originate. They are quickly metabolized but do not have the same effect on all tissues. In the penis, they relax the smooth muscle vascular tissue permitting blood to flow freely, hence their use in treating ED.

How Does This Medication Work?

Remember that it is blood flowing into the penis, and being trapped there during sexual arousal, that produces an erection. So getting "turned-on" relaxes the blood vessels in the penis and allows blood to flow freely into the corpus cavernosum. The action of filling-up with blood causes it to "pinch off" the out flow of blood, trapping it as long as you are sexually aroused—and this results in an erect hard penis.

Prostaglandin acts by immediately relaxing the blood vessels, even if they are getting only a weak or even no arousal signal at all from your brain and, of course, the corpus cavernosum fills up with blood and you have an erection. Amazingly, you could get an erection with this substance without being the least bit "turned-on"!

On the other hand, using this substance in a sexually arousing situation results in an *even firmer* erection! Ironically, these kinds of medications or substances were originally developed for application to other smooth muscle tissue, but their application to erection problems was soon realized.

How Can You Obtain the Medication Prostaglandin?

Once your doctor has given you a prescription, getting it filled is usually not a problem, just be advised that each prescription is prepared at the time it is ordered. This means that the pharmacist literally mixes ("compounds") the ingredients in the pharmacy. So it's best to have your physician call the prescription into your pharmacy ahead of time, or you'll need to be prepared for a short wait.

The cost of prostaglandin can be a concern for some users, but as Charles discovered, some health insurance plans cover the cost. Luckily for Charles, his health insurance covered the entire cost. Most of us may not be so fortunate, but be aware of the possibility that the cost *may* be covered by your health insurance plan.

What's involved in using an injected medication? Basically you need three things...

- Alcohol (to clean the area to be injected using a cotton swab or even a tissue).
- A 1-ml insulin syringe with 27-gauge needle (this could vary).
- A vial of the medication (usually 10 ml vial).

Sticking a needle in your penis to obtain an erection may sound intimidating to you, but let me assure you that dealing with the prick of the needle is easy. Some men feel no pain or discomfort at all.

For most men, the biggest hurdle they face is working this procedure into a romantic interlude with their wife! Once you understand the proper procedure for injecting into either the right or left cylindrical tube—the erectile body (the corpus cavernosum), you need only a strategy that will not interrupt the exchanges between the two of you. *One of the best approaches includes involving her as a participant in the whole process, perhaps even performing the injection.* It then becomes a part of the lovemaking.

Adventures In "Injection Land"

These injections have been used by thousands of men, and many strategies have been worked out to use this procedure privately. Here are a few stories you may find interesting and instructive...

Charles Uses It On His Wedding Night

Our friend Charles reports that his very first use of injection therapy was a big challenge to him. It was his wedding night—the first evening on his honeymoon in their hotel suite. He was prepared, having kept a small cooler bag with an ice pack until he could put it in the small refrigerator in the kitchen area.. He was determined to use this medication to assure that he did not have an erection failure with his new bride. *Anxiety is one of the biggest causes of erection failure,* and Charles was certainly fighting it off on this special evening.

As the evening wore on, his wife noticed that he seemed distracted, and probably to get his attention, she started to tease him seductively as they sat on the living room couch. Things quickly escalated to the hot and heavy stage and Charles was beginning to wonder if he was going to get an opportunity to use the injection. Fortunately for him, while nibbling at his ear, she whispered, "Meet me in the bedroom." Charles moved quickly to the bathroom adjoining the bedroom. Charles was now free to make his own dash to the little refrigerator where his precious "cargo" was stored.

He entered the kitchen smiling at his good fortune, thinking, "This is going to be easy." But, just before he started to inject, he realized he had forgotten to pack the alcohol he was supposed to use to clean the injection site and wipe the vial. With the wheels spinning in his head, he asked himself, "what else can I use?" His toiletry case happened to conveniently be lying on a chair in the living room, where he had left it right before her passionate approaches. He found a mouthwash with a high-alcohol content. It would have to do. And it did.

When he entered the bedroom, only a step ahead of her, amazingly enough, he had a rock-solid erection in spite of all his anxiety flutters. Feeling relieved, he thought to himself, "This stuff really works." In fact, some men who obtain the injection medicine report that the confidence it gives them is sometimes enough to assure them of an erection, whether they actually inject or not! They get an erection naturally, without injecting, because they know they have the medication to fall back on! We'll talk more about this later on in Chapter 10 — The "Mind Factor."

Harris's Daring Move

Another user of this medication is Harris, a sixty-year-old man with an erection problem caused by leakage of blood from the corpus cavernosum because of damaged blood vessels. This made keeping an erection very difficult for him, but he was greatly aided by the injection. He told me the story of how, on one occasion, he had arranged for his wife to meet him in his office one night while he was working alone.

In an adventurous mood, they decided to make love right there in his office. Right away, he thought of the perfect reasons to leave the office for a moment so that he could perform the injection. "Wait," he said as he went to his office mini-refrigerator and removed a thermos that belonged to the maintenance man and a brown bag containing the medication. "Fred, my janitor, uses my refrigerator. I am going to take his stuff down to him so he won't disturb us, and make a quick stop in the men's room." He quickly left the office, gave Fred his thermos, and turned into one of the washrooms, returning quite prepared with a ready erection!

Most husbands have no problem sharing that information with their wives. However, one who did was forty-five year old Mike. "I just couldn't think of a way to tell her," he told me. Although it might have been easier on him if he had told her, he nevertheless seems to have managed all right.

"I went to see my doctor because I was getting less and less hard. Sometimes it was difficult just to stuff it in. I was very happy when my doctor prescribed these injections. When Lil and I are going to have sex, I just go downstairs, get my secret little stash, and take care of business in the nearest bathroom. One day I will tell her, or she will find out on her own." It makes sense that the more open and honest your marriage relationship is, the more likely you are to share your feelings about your ED. You'll ask for your wife's support and understanding of the solution you've found, whether it be an injection or something else.

Some men ask, if the medication can be used well ahead of time. Unfortunately, it can't. The effect wears off in an hour or two and moreover, if the effect lasted more than four hours that would be cause for concern. An erection that won't go down requires emergency medical attention. Overall, however, this medication, if injected correctly, is considered safe, and most doctors will tell you that it has no side effects. The best indication of its safety is its approval (July 1995) by the U.S. Food and Drug Administration for use as a cure for impotence.

Careless injection can result in scarring, which can be a problem. A small number of cases have been reported in which scaring seems to have caused a curvature in the erect penis. One such case is recorded in the files of Dr. Bolton...

Jerry Finally Gets It Right

Jerry, a sixty-one year old engineer, reports that he knew something was wrong the moment he nervously injected himself incorrectly. He remembered thinking, "I didn't do that right," and he got only a weak erection. As time went on, he noticed the curvature in his erect penis and consulted his doctor who immediately took him off the medication. It was his doctor's opinion that he had injected himself incorrectly more than just once, contrary to Jerry's opinion. His story, though, has a happy ending. He learned to inject correctly, and after a few months, resumed using the medication with satisfactory results. For about two years now, since the scarring, he has had no problem.

There are more than 200 published studies on self-injection therapy. Adverse reactions associated with these injections have included mild or moderate penile pain, and much less frequently, priapism (prolonged erection). There is also a risk of scarring, as in Jerry's case, and hardening of the tissue, (fibrosis) especially with long-term use. A physician will probably not prescribe this therapy to men who have a predisposition to priapism or men with anatomical deformations of the penis and, of course, not to men for whom sexual activity is inadvisable.

What Are The Developing Trends in the Use of Injection Therapy?

Recently, these pharmaceuticals have been dispensed by some pharmacies in ready-to-use syringes, in dosages prescribed by the doctor. While this is a convenience to the user, since he does not have to fill the syringe from a vial, as described earlier, he does incur an increased cost. These ready-to-use syringes are generally more convenient but more expensive. Also they don't easily allow for varying the dosage, if the user would want to do that. So, in that sense, depending on your needs, the syringes may be less convenient. This is an important consideration for some users, with mild erection problems, who do not depend totally on the substance. Some users of the separate vial and syringe have reported that they adjust the dosage depending on how they feel. Sometimes they are able to achieve a satisfactory erection with less than a full dose or without using the substance at all!

Just the same, it must be noted that when a leading pharmaceutical company like Pharmacia & Upjohn, Inc. gains FDA approval and markets a product of this type, it must be recognized as a major gain for men with ED. Pharmacia & Upjohn's product, "Caverject," or its latest version "PenInject," is the pharmaceutical of choice for some doctors and their patients with ED problems. This pharmaceutical device, as you would expect, is reported to also contain prostaglandin, specifically referred to as alprostadil. Moreover, the Pharmacia & Upjohn company appears to have a United States-wide campaign underway to inform physicians and their patients who might need this therapy.

The Pharmacia & Upjohn delivery system certainly simplifies injection therapy. But there is speculation that, in the future, even simpler to use delivery systems will be developed. One can imagine, a patch, similar to the testosterone patch which could be painlessly applied to the penis as a way to administer this or a similar substance. Already, as noted earlier in the chapter, gels, which are either inserted into the penis or applied topically, are being developed.

Does The Idea of Using "Needles" Bother You?

Injecting the medication directly into the penis is a very reliable and efficient means to overcoming ED, and you are fortunate if it is an option that you can accept. Nevertheless, I know that there are men who have been using injection therapy quite successfully who, just the same, have been eagerly waiting, and hoping for something like Viagra. If they are among the 70% of ED sufferers who can be potentially helped by Viagra, it's a great day for them. Viagra is here and it works—in a more natural way! There is however, a trade off for those switching to Viagra. They will be giving up the instantaneous erection of injection therapy for the easier-to-take pill and a more natural erection. Viagra will not cause an erection *unless* the man is sexually stimulated.

Another significant revolution in applying prostaglandin for the treatment of ED has preceded Viagra by a few years and is already being used successfully by many men. In the next chapter, we'll take a look at urethral suppositories of alprostadil. It's a practically painless technique to induce an erection—and no needles are involved.

"Men with erection difficulties who were candidates for alprostadil, but who wished to avoid using a needle and syringe and couldn't use Viagra, welcomed this product. Most have reportedly been happy with the results."

Chapter 7

URETHRAL SUPPOSITORIES

"MUSE is another confirmation that 'potency-on-demand' is now a reality."

"Injection Therapy"...Without a Needle!

The urethral suppository technique was introduced in early 1996. Like injection therapy, it used a substance similar to prostaglandin, called alprostadil, to achieve an almost immediate erection. Why use it? Because *a needle is not required* to administer it. This is "injection therapy" without the needle and syringe.

The alprostadil is in the form of very small suppositories, which are inserted into the end of the urethra, the opening at the head of the penis through which urine and semen flow). When the alprostadil is deposited into that channel it will be absorbed through its lining and thereby initiate an erection, virtually the same as with injection therapy. In early 1997, doctors began introducing this product to their patients with ED. It is called MUSE and was developed by the pharmaceutical company Vivus, Inc. of Menlo Park, California.

A statement from their web page about the company and the product reads as follows. . .

> "MUSE (alprostadil) was cleared for marketing by the U.S. Food and Drug Administration in

> November 1996. Founded in 1991, Vivus, Inc. is a leader in the development of advanced therapeutic systems for the treatment of ED. Vivus has pioneered a novel therapy for ED known as the transurethral system for erection. This therapy consists of a proprietary, non-invasive, drug delivery system that delivers pharmacologic agents via the urethra."

As with injection therapy, the goal is to get prostaglandin, or in this case alprostadil, into those two spongy bodies (the corpus cavernosum). There it will relax the arteries allowing blood to fill them and thereby produce an erection. In this medical technology breakthrough from VIVUS, the alprostadil is absorbed into the corpus cavernosum from the urethral channel.

Here is an explanation from VIVUS:

> "MUSE (alprostadil) consists of a discreet, single-use, pre-filled, plastic applicator containing a very tiny suppository of the drug alprostadil, a medication which increases penile blood flow. The stem of the applicator is inserted approximately one inch into the urethra where the medication is released, initiating an erection within five to ten minutes, which typically lasts thirty to sixty minutes. Transurethral therapy is a departure from existing therapies for ED and is based on the discovery that certain medications can be absorbed by the urethra and transported into the surrounding erectile tissues."

Men with erection difficulties who were candidates for alprostadil, but who wished to avoid using a needle and syringe and couldn't use Viagra, welcomed this product. Most have reportedly been happy with the results. If there has been any disillusionment, it has been with the cost. At the present time, this system is considerably greater in cost than most other "cures for impotence." It is likely however that as the system becomes more widely used, the cost will decrease.

How MUSE Works

To begin with, before you receive a prescription, you will have used this system in your doctor's office. Once it has been determined that this therapy is appropriate for you, you will receive a prescription especially formulated in the dosage that your doctor has determined will meet your needs. After the pharmacy has filled your prescription, you will have packets that contain a single applicator with a single dose. The applicator is essentially a thin stem which is small enough to insert into the penis opening, the urethra. To this stem is attached a small device with a button for releasing the suppositories. The packets are to be kept under refrigeration, and one should not be opened until you are ready to use it.

Before inserting the stem to release the medication it is advisable to urinate and shake down the penis, to remove the excess urine, then gently stretch the penis to its full length. If any urine remains, it will be helpful in dissolving the suppositories. After inserting the stem about one inch into the urethra and pressing the button to release the suppositories, remove the device and gently roll the penis between the palm of your hands to assure distribution and absorption of the medication. Also, your doctor will give you other advice to make the medication as effective as possible for you.

MUSE is yet again another confirmation that "potency-on-demand" is now a reality. And the side effects are minimal—in fact, only a few men experienced any. The side effects however, include: aching in the penis, testicles and legs, in the area between the penis and rectum, warmth or burning sensation in the urethra, redness of the penis due to increased blood flow and minor urethral bleeding or spotting due to improper administration. There are probably other less reported side effects about which your own personal physician may determine that you need to know based on your health history or other factors.

Some men express concern about inserting an object into the urethra. But they can be reassured by the fact that every day thousands of men have tubes of greater diameter (catheters) inserted into their urethra. So inserting a small tube into your urethra is “no big deal.”

Is the resulting erection as satisfactory as that from injections with a needle and syringe? The “jury” is still out on that question, and besides, the resulting erection depends on the severity of your ED. Experience seems to indicate that even if you increase the volume or strength of the dosage, the resulting erection still depends on the severity of your ED. Regardless, the bottom line is that you will almost assuredly achieve an erection. As one satisfied user said, “This stuff works!”

There are reports that some men switching from injection therapy found that urethral suppositories were not as effective as injections of prostin directly into the corpus cavernosum with a needle and syringe. Just the same, some men using MUSE are convinced they have found the right solution. A particular example stands out: Eighty year old Ben reports that his hands were too unsteady for him to feel comfortable with a “needle,” but he had no trouble inserting a urethral suppository which, as he reports, “works just fine.” Younger men, as you might imagine, are even more enthusiastic.

Certainly MUSE promises to be a great triumph over ED and, for many men, a welcome improvement over injection therapy, albeit, at this time, a rather expensive one.

“What a man wants to do he generally can do, (when) he wants it bad enough.”

Louis L’Amour

Chapter 8

VACUUM THERAPY

"An erection achieved with this appliance is truly reliable. Not even premature ejaculation can slow you down."

"I Feel Like A Man Again!"

If you can't or won't use VIAGRA, perhaps you'll opt to "pump-it-up." Men, who have not had an erection in years have had their sex life returned to them through vacuum therapy. As one happy discoverer of vacuum therapy put it, "I have been given back my manhood." Vacuum therapy is easy and effective and the downside is minimal. It is also relatively inexpensive but, be wary of "bargains."

Shortly before VIAGRA came out, it was praised by a happy user whose letter was published in Ann Landers' syndicated advice column. "Don't Let Impotence Destroy Your Sex Life" was the title for that day. The praise of this writer was certainly justified. "I feel like a man again," he wrote, "Imagine having a strong hard erection 'on-demand.'" That is virtually the promise of vacuum therapy. Impotence is overcome, and mild or even severe erection problems give way to strong, sturdy erections.

What is Vacuum Therapy?

- The number one thing to keep in mind is that it is self-administered—you do it yourself.

- Secondly, you must have a specially designed vacuum pump, which is available by prescription.
- And thirdly, it is non-intrusive. It requires no injections, no suppositories and no surgery, *and does not involve taking any medication.*

Vacuum therapy uses physical principles similar to those already at work in attaining a natural erection. In a normal erection, blood is pumped into the penis by the natural and ordinary working of the circulatory system on signal from a complex array of arousal impulses via the nervous system. Actually the most crucial step in maintaining an erection is keeping blood from flowing out of the penis—once it is engorged. Therefore, the sexual arousal signal going through the nervous system ultimately results in restricting the flow of blood back out of the penis once it has entered.

In vacuum therapy, by means of a clear plastic cylinder placed around the penis to which is attached a small pump, a vacuum is created around the penis. Normal air pressure around the penis is reduced causing blood to flow freely into the penis. Miraculously, it seems, a very firm erection emerges in the cylinder—right before your eyes. And you, the user, are in control every step of the way. The crucial step of maintaining the erection is achieved by easing a flexible ring off the end of the cylinder so that it gently constricts the penis at its base. Like a tight rubber band, (ordinary "rubber bands" are not recommended!), it prevents the blood from flowing back out of the penis.

This method of achieving an erection has been certified by doctors to be completely safe, and a major manufacturer of the leading vacuum appliance claims that no one has ever been injured using its product. Indeed, it is hard to imagine any side effects at all as a result of using the device, if used according to instructions. However there is one precaution users are urged to take: you should remove the flexible ring after intercourse—and, in any case, you must remove it after 30 minutes. If you plan to go longer, you may apply the vacuum pump and put it back on after 15 minutes.

An erection achieved with this appliance is truly reliable—not even premature ejaculation can slow you down. In fact, many men feel they have overcome premature ejaculation through its use. Still, others tell of lengthening their penises through regular use of the appliance. It is quite likely that if your penis is not already at its full length potential, using the pump will get it there. One of Dr. Bolton's patients illustrates the kind of success men have using this device...

Bill Overcame His Shyness And Regained His Potency

Bill is a high school teacher, who at thirty years old, was still, believe it or not working on overcoming shyness. He is a pleasant person with ordinary but good masculine features. In every way he was a very "straight-arrow" kind of a guy. He was successful in his profession, but incredibly his shyness caused him to have no confidence in having sex with his wife. He met Dr. Bolton through a self-help group he had joined to develop self-assertiveness.

After much prodding, Dr. Bolton was able to get through his shyness to discover that his problem was that of losing his erection just at or shortly before the crucial moment of insertion to have intercourse. During his teenage years and early adulthood he described himself as being obsessed with sex because, it was continually on his mind. "It was all I could do," he said, "to keep from walking around with a hard-on all the time." He went on to say that the few times he and his wife had sex were disastrous. "Sometimes, I would ejaculate while I was trying to get it in, and at the same time I would be losing my erection." Even so, his wife loved him very much and was being very patient and understanding with him.

Dr. Bolton recognized that the conventional way of treating Bill would have been to give him extensive counseling, psychoanalysis, or a combination of both. Nevertheless he decided on a short-term solution, reserving his more conventional recommendation of extensive counseling possibly for a later time. Dr. Bolton recommended vacuum therapy.

Bill was referred to a urologist who, after a brief consultation and exam, gave him a prescription for the appliance. Bill was delighted to discover that his health insurance coverage paid most of the cost. Keep in mind that good ones are not cheap and cheap ones aren't worth buying.

Something suddenly occurred to him, "How am I going to use this thing when I'm with my wife and we're "heating up" to the point where I'll need it?" Thinking about it made him nervous, but he knew this device could produce an erection whether he was nervous or not.

But as the saying goes, "Love will find a way." And it did—at first, his wife didn't even notice the small ring at the base of his penis. When she finally did, he casually told her that "It keeps me hard longer." Otherwise, she seemed too pleased to care. As time went on, Bill discovered that he didn't need to use the ring to maintain his erection—simply pumping it up to get started was enough. For many men with mild erection problems, achieving an initial erection with the pump may be all that is needed, particularly in a sexually stimulating situation. They find they are able to remove the flexible ring and still continue with their lovemaking.

The Case Of Frank Illustrates The Wide Applicability Of This Appliance

Frank, a fifty-eight year old businessman with a number of health problems, whose doctor wouldn't prescribe VIAGRA, hadn't had an erection in eight years. It was his wife who urged him to see a urologist. When he did, he was surprised to learn that his case might not be hopeless. He was offered injections but, like a lot of men, the idea of needles made him uneasy. However, he readily accepted the idea of vacuum therapy. He particularly liked the fact that he would not have to take any other medication, besides what he already took, for his other health problems.

Understandably, both he and his wife were pleased with the results. Using the device was no problem for Frank, as his wife was a supportive, willing participant in its use. It became a part of their lovemaking. The return of Frank's sexual functioning brought romance back into this couple's life. And even though they had been married for more than thirty years, they felt like newlyweds!

Some Finer Points About Using Vacuum Therapy

Jim was very excited about trying vacuum therapy. He was hopeful that it would enable him to overcome his mild ED. His complaint to his doctor had gone something like this: "Doc, sometimes, it just doesn't get hard enough–I can just barely get it in." His doctor reassured him that this was not an uncommon complaint for a man of 59 years. Nevertheless, he prescribed vacuum therapy.

When Jim tried to use the device on his own, he found it painful. It took quite a bit of practice and consultation with the manufacturer before he became adept at using it. The problem Jim was having was that he was allowing scrotum skin and other surrounding skin tissue to be suctioned into the tube surrounding his penis. This, in itself, can be uncomfortable. But he was also releasing the constrictor band in a way that unnecessarily trapped blood under the skin at the base of his penis, causing little raised areas that looked like tiny balloons. The solution was to quickly adjust the constrictor band and massage the "balloons" away as he was removing the tube.

Jim learned that if he pressed the tube firmly against his pubic area as the vacuum was being created, he was less likely to suction skin tissue into the tube. He also learned the technique of pointing the tube slightly upward as he began introducing the vacuum. Then, after a moment of giving the vacuum enough time to have an initial impact, he would point it downward or otherwise adjust it to allow blood to flow more freely into the penis.

Jim also discovered it is best not to rush the process. It doesn't take a lot of time, but slowing down the process and applying the vacuum pressure gradually allows the slowly growing penis itself to help prevent the surrounding pubic and scrotum skin from being suctioned into the tube. Finally, Jim found that it was best to choose a tube no bigger than he really needed, no matter how big he thought his penis was! When he mastered these finer points of applying vacuum therapy, the whole process became a snap with no discomfort to speak of.

How To Obtain And Use A Vacuum Therapy Appliance

This therapy has been available since the early 1980's. At the time of its introduction, urologists were skeptical, but soon found their patients accepted the devices, and that they preferred them to other options they were being offered. This therapy for treating ED was first developed by Geddings Osbon, Sr. during the 1970's. The process is, in fact, now referred to as The Osbon Technique. His development of the technique was actually in response to his own ED problems. Mr. Osbon was determined that he wouldn't allow his erection difficulties to prevent him from enjoying the sexual experience.

An appliance is available today from a company bearing his name, Osbon Medical Systems, marketing a product called the ErecAid System. Information about this company and its products are widely available, for example, on the worldwide web. This was the first company to receive approval (1982) under FDA guidelines for the manufacture of a vacuum device for "impotence management."

Today there are several manufacturers of similar devices. They include Pos-T-Vac of Dodge City, Kansas, VET-CO, Inc.(Vacuum Erection Technologies) of Birmingham, Alabama and some others. Most of these companies offer several models including manual hand pumps and battery powered pumps. The battery-powered pumps may be helpful to men with limitations on the use of their hands such as might be the case of a user with arthritis or other debilitating conditions.

You can obtain an effective vacuum appliance through a medical doctor—most likely a urologist. He can give you sound and helpful advice and you can rest assured that the advice you are receiving is from a member of the medical profession, through which you will receive a warranty and support.

Once you have the device, experiment with its use. You will

be supplied with a full set of instructions, and possibly an instructional video, by the manufacturer or your doctor. But here are a few hints to keep in mind: be sure to choose the right size constrictor ring and adapter sleeve, which is used to adjust the opening in the cylinder to fit your penis. Some manufacturers make adjustment for size by supplying several sizes of cylinders. If the adapter or the cylinder is too large, excess skin from around the scrotum will be suctioned into the tube. If the constrictor ring is too small, it will be uncomfortably tight and restrict blood flow more than necessary. If this ring is too large, it will not retain enough blood in the penis to maintain an erection.

Finally, it's common for some men who wish to enhance their erection to have a variety of tools at their disposal; the use of vacuum therapy will be just one among several means of achieving their goal. Recalling Larry, the inspiration for this book, he too obtained a vacuum therapy appliance. But, as you may recall, he used other approaches as well and was particularly pleased with injections. It was his practice to alternate between the two therapies. On the other hand, there were times when taking the yohimbe pills, alone, was sufficient to enable him to achieve a satisfactory erection on his own.

I also know of one individual who regularly uses injection combined with vacuum therapy. He said that, on one occasion, he decided that he wasn't satisfied with the erection he had achieved through injection and it occurred to him to improve it with his vacuum device. He did, and happily reported that the result was fantastic. Certainly caution would be advised in a case like this, to avoid injury. Applying the same amount of vacuum pressure around a partially erect penis, as you would a flaccid one, to achieve an erection, could result in bleeding.

It's imperative that you exercise good judgment. That coupled with professional advice, will be your best guide if you are inclined to apply a variety of ways of enhancing your erections. *Manufacturers of pharmaceuticals are quick to advise against using any combined methods in the treating of ED. No clinical studies have been done using this approach.*

"We accept conditioning our minds for mental tasks and our bodies for physical work and athletic competition. Why not also accept conditioning the penis for the performance we want from it? If you think about it, you will probably agree that anything we do with our bodies can be improved through conditioning."

Chapter 9

Conditioning The Penis

"You will soon notice that your erections are improving by how quickly you attain them and by their firmness and duration."

Use It Or Lose It!

Most men take their penises for granted; yet they expect them to perform ably when they want them to. We accept conditioning our minds for mental tasks and our bodies for physical work and athletic competition. Why not also accept conditioning the penis for the performance we want from it? If you think about it, you will probably agree that anything we do with our bodies can be improved through conditioning. You can condition your fingers to be a better word processor, your eyes to be a faster reader, your ears to make finer distinctions in sounds. Likewise, nonuse of any skill or ability can result in atrophy or loss of skill or ability. The old adage of "Use it or lose it" seems to be universally true—there is a lot of agreement that it applies here too. Here then is a conditioning therapy for the penis.

Heat Therapy

A more appropriate name for this treatment is "hot and cold water" therapy. It is based on the fact that an erection is achieved when the penis is engorged with blood. The physiological body principal is that blood flows more freely through or into a body part that is heating up, and therefore needs cooling down and retreats from one that is getting too cold.

Basically, It Consists Of Alternately Warming And Cooling The Penis

This therapy needs to be practiced over an extended period of time, six-to-ten weeks, in order to derive maximum benefit. It is especially recommended for its rejuvenating effect. Dr. Bolton indicates that he often recommends it to men coming out of depression who have been sexually inactive for long periods of time.

But any man can benefit from it, and the results are often very impressive. For example, one man who regularly uses this therapy reports that he steps out of his shower, where he performs the therapy, with a strong ready erection and heads for the bedroom.

What This Therapy Is All About

This is a therapy best performed in the privacy of your bathtub or shower. Once you understand the principles, you will be able to devise the steps best suited to your situation. Since you will be dealing with hot and cold water, a word of caution is in order: *you must be sure that your source of hot water is not so hot that it could injure you.* Here is what you will be attempting to do…

- Introduce heat to your penis by wrapping and holding it in a hot wet cloth for three or four minutes,
- Then suddenly introduce it to very cold water while it's still wrapped in the cloth, for another three or four minutes.
- Repeat the above procedure, four-or-five times, always ending with the heat treatment.

How To Perform Warm And Cold Water Therapy

Here are two ways to perform this therapy. You will be able to adapt these ideas to fit your situation.

Method One is done in your bathtub. Sitting in water in a bathtub, have a source of hot and cold water nearby—like a kettle of hot water and a container of cold water. If your tub is equipped with a hand-held shower and a temperature sensitive single-valve control that can quickly switch from hot to cold, you could use that instead.

You will need an absorbent cotton cloth to wrap around your penis several times. You could use towels or strips of absorbent cotton such as the cloth used for baby diapers. Cheesecloth also works very well.

Wet the cloth around your penis with room temperature water first, then slowly pour or spray-on the hot water—as hot as you can stand it. When the cloth is saturated and your penis is very warm, wait several minutes, pouring-on more hot water as needed. Soon, you will start to feel your penis enlarge. Next, start to pour very cold water onto the cloth. You may wish to avoid letting the cold water contact any other part of your body, other than the penis, to avoid the shock of the cold.

The coldness will start to make your penis retract—that is expected. But just as your penis retracts a small amount, quickly begin again to pour or spray more hot water onto the cloth that surrounds it. As the heat reaches your penis it will start to enlarge again.

Repeat this process several times being careful not to scald yourself. Always end with hot water. You will very likely step from the tub with an erection. And don't be surprised if you also feel very refreshed and exhilarated too.

Method Two is not that much different from method one. It requires that you have a shower equipped with a single handle control so that you can go from hot to cold water by turning the handle from left-to-right or right-to-left. Practice changing from hot-to-cold before you start your treatment.

Since this treatment is performed in the shower, you begin by standing in the shower as you would ordinarily. Wrap your penis with the cotton cloth as described above.

Begin to slowly saturate the cloth with warm water from the hand-held showerhead, then slowly turn the valve toward hot until the temperature reaches as hot as you can stand. Continue letting the water run for several moments, then reverse the direction of the valve turning toward cold, until the cloth around your penis is saturated with cold water.

Keep it cold for several moments, then reverse again, this time toward hot. Do this until the water is again as hot as you can stand it, and keep it hot for several moments.

Repeat the entire procedure several times, and always end with hot water. As in method one, you will probably step out of the shower with a firm erection and feeling great.

For best results you should perform this therapy over a period of at least six weeks and for at least three times a week. Don't be discouraged if a dramatic improvement doesn't result from the first time you use this therapy. You will soon notice however, that your erections are improving by how quickly you attain them, as well as their firmness and duration.

Chapter 10

THE "MIND FACTOR"

"Since the human body tends to move in the direction of its expectations, plus or minus, it's important to know that attitudes of confidence and determination are no less part of the treatment program than medical science and technology."
Norman Cousins

Your Mind Is The Ultimate Sex Organ

It may seem hard to understand that, while you are desiring to have an erection, it is possible that something in your own mental machinery may be working against you. Nevertheless, because of the peculiar complexity of the human mind, it's true. Ordinary operations of the mind, like forgetting and remembering, illustrate this complexity.

Everyone has had the experience of trying to remember something, only to have it reappear in the mind, i.e., finally remembered. Where had it gone? Does the mind have a record of all things you've ever experienced? And what about dreams? Where do they come from? These simple, yet profound, questions are the basis for the theory that forgotten experiences can affect our lives, our bodies and yes, even our erections!

Forgotten experiences reside in the realm of the subconscious part of the mind, sometimes called the *unconscious,* and only rarely appear in the conscious mind. We will use the term *"subconscious"* rather than *"unconscious,"* following the practice of many other professionals in mental science who believe that *"subconscious"* is the more correct terminology. The "subconscious" refers to that part of your mind that is generally out of reach, but with which you will learn to communicate in order to enhance your erections.

Forgotten experiences influence our emotions, behavior and physical condition. They can also influence our sexuality, including the ability to achieve an erection! Many in the field of psychology and psychiatry have devoted their careers to the exploration of this phenomenon. But here, in brief, is what you need to know…

Can Psychological Factors Affect Your Ability To Have An Erection?

Consider this chapter a footnote. Why? Because the methods already outlined will (in most cases) give you the erection you need—even if your subconscious mind is working against you. Just the same, it would be better to remove a psychological block rather than just to override it. Furthermore, this same mind power that may be working against you can be harnessed to work for you. *You can improve your erections and your sexuality through the power of your mind by programming your subconscious to accept and act on your true desires.* So, even if you don't have a psychological block, you can use your subconscious mind to have better erections!

If you have problems attaining a satisfactory erection, determining whether or not the problem is resulting from influences at the subconscious level can be tricky. But, in most cases, if you can attain an erection in some situations but not in others, then there is a significant probability that the problem is mental. In any case, you don't have to be sure your erection difficulties are mental to benefit from programming or reprogramming your subconscious. You are very likely to benefit from a therapy that programs your subconscious to meet your desires, whether there is an existing problem or not.

An example case of a man with an actual problem is recorded in the files of Dr. Bolton—we'll call him Jack. Jack was seeking help for what he described as persistent impotence. However, Dr. Bolton learned, in the course of the therapy, that Jack regularly had wet dreams and often woke-up with a hard-on. But his erections nearly always failed him when he attempted intercourse. After only a brief period of analysis and hypnotherapy, Dr. Bolton concluded that Jack's problem was at the subconscious level. He was responding to conditioning by his parents induced during his childhood. So effective were his parents in condemning his sexuality that, at the most crucial moments in his adult sex life, his erections just faded away.

It got worse —as time went on, anxiety and worry seemed to make failure unavoidable—clearly a case of "performance anxiety". Even though Jack was happily married, he was unable to perform with his wife. Dr. Bolton was able to help him break through the repression of his sexuality by going directly to his subconscious mind, through the use of positive affirmations. These are simply statements of how you want things to be, as if they already are that way! Many challenges, sexual or otherwise, have been overcome in this way.

What Are Positive Affirmations And How Do They Work?

One of the purposes of the subconscious mind is the preservation of its owner. However, once an individual has accepted a belief or an idea thoroughly enough, it becomes entrenched in his subconscious. Even though the idea may be harmful to his well-being, the subconscious mind believes *any* idea that's delivered to it with great emphasis. It doesn't matter if the idea is in the individual's best interest or not. It simply drives the person to do what is necessary to make it happen!

A belief, or an idea, entrenched in the subconscious mind, can continue to influence a person's actions long after he has forgotten all about it, as in Jack's case. Through positive affirmations, the therapist uncovers and endeavors to replace the harmful belief with a helpful one. The process of doing this is neither magical nor radical.

There are two important steps involved...

- First the conscious mind must be quieted. The conscious mind is the "I" part of you that is continuously engaged in thinking, or "self-talk," as you go about your waking hours.

- Secondly, once the conscious mind is quieted down, communication with the subconscious mind can take place. The therapist or hypnotist (or you, yourself) attempts to quiet the patient's conscious mind, in order to bypass it, so communication may be established directly with the subconscious. There are various techniques for quieting the conscious mind, but the idea of someone waving a watch before the face of a subject and counting, as in hypnosis, while suggesting that the subject is becoming sleepy, is essentially correct. The person under hypnosis is not really asleep, but his conscious attention is focused or directed on a single object. With the conscious mind relaxed or focused, the hypnotist is able to "by-pass" it and implant suggestions directly into the subconscious. This is how the subconscious mind can be "programmed," or perhaps "*re*-programmed," with helpful beliefs.

The memories, or experiences, entrenched in the subconscious that interfere with proper sexual functioning, can also be removed, reinterpreted or replaced. And, since the subconscious is such a powerful "influencer" of our bodies and our lives, it is possible, through it, to change or improve conditions in our sex life. Even if an erection problem is not the result of repressed memories or forgotten experiences, it is possible to reprogram the subconscious to cause improvements in the body, including better erections.

The key to tapping into the subconscious is to displace or remove the influence of your everyday regular thinking mind—

the "I" that is doing the continual self-talk. You can't displace it entirely, nor would you want to. But quieting all its busy thought activity to a drowsy state by focusing its attention to a single point—a single monotonous item like a candle flame or vase—is all that is necessary. Amazingly, this is something you can do yourself. Self-hypnosis is not only possible but it is practical and easy.

How Can You Overcome Erection Problems Through Positive Affirmations? What Can You Do About The Psychological Factors Affecting Your Erections?

The amazing thing about affirmations is that you can do them yourself. The process of communicating with your subconscious mind does not require a third party. In fact, even under hypnosis induced by an expert hypnotist, it is entirely up to you whether or not you accept the suggestions he gives. So, not surprisingly, there are techniques you yourself can use to quiet your own conscious mind without assistance from anybody. Yes. You can program or reprogram your subconscious mind.

Using Positive Affirmations To Improve Your Erections

If you are seeking to improve your erections, you can have the added benefit of making sure that your body and your psyche understand your intentions. Simply harness the power of your subconscious mind by making positive affirmations. Just as the body tends to heal itself, though often not without help, the natural functions of the body can also adjust or heal themselves. The complex set of bodily events involved in producing an erection can be influenced by your subconscious desire. In other words you can program your subconscious mind to help give you better erections. Your communications to your subconscious mind must be expressed in positive, affirmative terms and must be believed in by you. For example, you could read this aloud to yourself every day…

"My body is functioning in a healthy and normal way. I achieve a strong, hard erection whenever I am in a lovemaking or sexual situation. I am able to share great pleasure with my mate."

Here's How You Can Achieve a Hypnotic State On Your Own (Self-Hypnosis)

- Choose a quiet dimly lit place where you will not be interrupted.
- Sit or lie quietly in a relaxed state.

Sit quietly in a comfortable chair or lie quietly on a mat, or in bed and let your body sink heavily into it. Focus your attention first on your right arm. Raise it slightly, then imagine that it is very heavy. Let it drop heavily as if it were not in your control. Do the same with your left arm. Turn your attention to your legs. Raise your right leg slightly and experience its heaviness, then let it drop heavily as you did your arms, and feel it relax. Do the same with your left leg. Also, let your head relax in the same way. Finally feel your entire body sink into the chair or mat very relaxed. Spend a few moments enjoying this relaxed state.

Now, focus on a lit candle or other object, such as a rose in a vase. Place it in your line of sight so that you can see it without moving or straining. The room should be dimly lit. Begin slowly to count backward from 20. All the while, look intently at the flame of the burning candle or whatever object you have chosen. Think of nothing else. If you find other thoughts entering your mind, don't be concerned—just dismiss them at once and continue counting. Other thoughts will enter your mind, but let them go as quickly as you can. Continue this process for as long as thirty minutes. The whole process can be completed in as little time as ten minutes—maybe less.

When you find yourself very relaxed, possibly to the point of drowsiness, begin to repeat affirmative statements, such as...

"I am relaxed and confident. I feel myself free to be the person I choose to be. I am confident about my sexuality. My body functions in harmony with my desires with ever more vigor and stamina. My erections come freely and with certainty in every sexual situation. They are hard and strong. I see myself performing vigorously, giving and receiving great pleasure and joy."

To achieve a deeper hypnotic state, it may also be helpful to imagine yourself in a situation or place that you consider serene and peaceful. This could be a meadow in springtime, where wild flowers are blooming profusely or in any natural setting you enjoy. As you count backward, see yourself going down a gradually descending path in a comfortable and pleasant place. Or down a winding staircase from a beautiful castle-like building to the same pleasant meadow.

Once you feel you have arrived at the deep place in your mind, almost drowsy and oblivious to your surroundings and you feel pleasant and relaxed, then state over-and-over your affirmative statements-your positive definition of the true self you have elected to affirm as just described. (See the affirmation of two paragraphs ago.)

You could improve your erections through self-hypnosis. When you take care to safeguard your health and promote your vitality, the improvement can be dramatic. I have included only the basics of what you need to know about self-hypnosis just so you are aware of it. Therefore, discussion of the subject and its possibilities are much abbreviated. Should you wish to delve deeper into it, you may want to seek out the help of a qualified mental health professional.

Where Can You Learn More About The "Mind Factor"?

There are many self-help books, like *The Power of Positive Thinking* by Norman Vincent Peale and also audiotapes that can help you build confidence in yourself and your ability to get what you want. It may be helpful for you to explore some of them on your own. The more thoroughly you understand this process, which is part of your overall success, the better you will be able to make it work for you. Some of the basics are included in this book but, I repeat, you are encouraged to explore the subject further.

Here are some more personal development books that can help you. All the principles in these fine books work together and contribute to your overall well-being, and thus could affect your sexual functioning.

- *No Excuse—I'm Doing It!* by Jay Rifenbary
- *Reject Me—I Love It!* by John Fuhrman
- *Brighten Your Day With Self-Esteem* by William McGrane
- *What to Say When You Talk to Yourself* by Shad Helmstetter
- *The Seven Habits of Highly Effective People* by Stephen R. Covey
- *Psycho-Cybernetics* by Maxwell Maltz MD, F.I.C.S.
- *The Greatest Secret in the World* by Og Mandino

The books just listed have the potential to support you tremendously in molding your thinking, to help you achieve the life you really want. Internalizing the wisdom of these authors could result in a mental fitness and wisdom such that each day of your life would be a day in which all of your unique gifts would be expressed to their fullest. You might not then need actual hypnosis such as I have described. But the effectiveness of actual hypnosis in overcoming ED in cases in which the cause is entrenched in the "mind factor" is well confirmed. Therefore here are the titles of two books from Dr. Bolton's list. He keeps copies of them in his office for use by some patients, as he recommends. The following publications may only be available in libraries, but check them out, if you can. Then, look for similar books in your bookstore:

- *Helping Yourself With Self-Hypnosis*, by Frank S. Caprio, M.D., and Joseph R. Berger
- *Healing With Mind Power: Living and Feeling the Way You Want To-Through Guided Meditation and Self-Hypnosis*, by Richard Shames, M.D. and Chuck Sterin, M.S., PhD

I also recommend that you check out the availability of courses in hypnosis or self-hypnosis that may be available in your area.

I cannot resist pointing out that all the books listed in this section, either imply or state directly that a positive focus is essential to bringing about positive results. Some call it the "law of attraction." You will attract to yourself the joy and success that you believe you deserve. The triumph of Viagra is permitting men to overcome limitations and I believe it is the nature of human kind to attempt to triumph over every kind of limitation.

In dealing with what we are calling the "mind factor" we are facing a subtle assault on our potential. Overcoming "mind factor" problems may call for a thorough examination of yourself. As you progress, and perservere, you will be enlightened by the fact that once you believe in joy and success for yourself, you are likely to believe in it for others as well.

A positive focus, maintaining an expectation that you are achieving your goal, is crucially important. All though there are medications similar to Viagra being developed that may one day enable a man to overcome ED even if his subconscious mind is working against him, the potential to overcome "mind factor" problems on your own is within you. Focusing on your goals and persistently experimenting with solutions is the key to every successful triumph.

"The greatest discovery of my generation is that man can alter his life simply by altering his attitude of mind."
William James

"...even though about 30% of men with ED cannot be helped by Viagra, they need not be discouraged. In all likelihood, they can use one or more of the other methods, as described in this book. There's probably a solution for almost any ED problem there is."

Chapter 11

YES—ALL MEN CAN NOW HAVE HOPE OF OVERCOMING ED!

"With the help of your doctor, what's best for you in overcoming ED can now be determined."

Hope Abounds As New Advances Are Made

Though millions of men with ED, in the U.S. alone, will not or cannot be helped by Viagra, there's hope for them too. Certainly, only a few physical impairments could stand in your way of overcoming ED and, as you have learned, there are several approaches to enhancing or improving your erections. New developments toward overcoming ED are appearing with amazing rapidity. In the November 17th, 1997 issue of *Newsweek* magazine, the subject of exploring advances in the "cure" of impotence was covered extensively, if not a bit sardonically, as it proclaimed "the brave new erotic world, filled with chemical promise...."[7] This book has been examining the promises and the prospects that men, and the women who love them, will be less and less hampered by physiological failures of the body.

There can be little doubt that the stress and challenges of modern civilization bear much of the blame for the loss of someone's ability to enjoy the physical expression of love between the sexes. Also, there can be no doubt that the needs and urges of sexual expression are almost undeniable among human beings. Experts in science and medicine are demonstrating that they can overcome the damage we have incurred. With the information in this book and the help of your doctor, the best solution for overcoming ED can be determined. You now have the information you need to greatly improve your sex life and that of the one you love.

Now that Viagra, a pill that effectively enables the majority of men to overcome ED, is available, many expect it to become the treatment of choice. As mentioned earlier, it is reported to be ineffective in 30% of the men who could use it. In addition to that, there are many men who should not take Viagra either because of their health or other drugs that may cause problems when mixed with it. That means that probably at least 9 million men with ED, in the U. S. alone, will have to seek another treatment. Hopefully, this book can help them.

It must be clear to you now that the answer to Larry's question, "Isn't there something for guys like me who just want to get it harder?" is a resounding YES! More importantly, men with serious ED who might otherwise have had no hope at all of ever having an erection again, now have it in their power to have their potency virtually restored. They can enable themselves to engage in sexual intercourse, if they so desire. If this information is judiciously applied, it can enhance or return your erections and thereby enhance your love life. You need add only your determination and a little self-discipline. Here are a few more examples of men who have done just that. They have squarely faced their erection difficulties and overcome them:

How Sam Regained His Potency After His Prostate Was Removed

Sam was a young fifty-six years old with his libido totally intact. The problem was, he had undergone a radical prostatectomy in order to cure him of the cancer his urologist had detected during a routine physical examination. In this operation, his prostate gland is completely removed. As you may know, the nerve bundles that carry the signal from the brain for causing an erection, are located on each side of the small walnut-sized prostate.

In cases in which the malignancy is not near the nerve bundles, the surgeon, during the operation to remove the prostate, can spare or save the nerves. In Sam's case, unfortunately, because of the location of the malignancy, it was not possible to spare the nerves and they were lost or damaged. Just the same, Sam emerged from the operation grateful to be pronounced "cured" of cancer and otherwise as fit and healthy as he had ever been. He had even quickly recovered from the "to be expected" incontinence. Nevertheless, his inability to have erections bothered him, even though he had been well-advised that it too was "to be expected."

It was very disconcerting to Sam to be in intimate closeness with his wife and have absolutely nothing happen with his penis. Well, not quite "absolutely nothing"—in his case, he did feel a pleasant "stirring." Even though there was never any doubt that Sam's libido was working just fine, his doctor had counseled him to be prepared to deal with the loss of his erectile functioning. So in a manner of speaking, he and Hilda (his wife) were prepared. They accepted that a bit of planning and anticipation would necessarily always accompany their love life in the future.

Sam had resolved in his own mind that he would continue to have sex with his wife as often as he had always done. Within a few months after his surgery he had learned to use injections. The couple was pleasantly surprised at how well they worked. "As well as before," Sam said. "Better," Hilda replied. In fact he had to admit to himself that he was even getting better erections than he had previously gotten naturally on his own!

As far as Sam was concerned, there was just one drawback. There was no way was he going to inject every day. Besides which, his doctor had put definite limits on how often he might use this therapy. For Sam that seemed to mean three-or-four times a week. But he soon became concerned about injecting even that often. When he discussed his concerns with his doctor, the doctor, without hesitation, introduced him to the idea of vacuum therapy. "No injections, no drugs," said the doctor.

Sam could hardly suppress his amusement as he examined what the doctor had placed before him. Could his doctor be serious? This looked like something you might run into in a porno shop—the idea of "pumping up your penis." "What a hokey idea," he thought.

But he tried it, and it worked. Both he and Hilda were delighted. Although he had lost his prostate gland, he had been cured of cancer, and he was still able to carry on a very satisfactory sexual relationship with his wife. Best of all, he and his wife agreed that, his erections were quite satisfactory.

In some cases, erections achieved through vacuum therapy have a tendency to have a "hinged-like" effect. The erect penis does not stand out straight, but swings downward from the base of the pubic area at the point that the constrictor band is applied. But this was not the case with Sam. His was a "stand-up" erection—a welcomed outcome, attributed both by him and his doctor to a well-developed "PC" muscle (See the explanation of Kegel exercises in Chapter 2.). Luckily, Sam had performed Kegel exercises on a regular basis prior to his surgery. Quite likely, Sam will always require some form of erection treatment to attain an erection, but he has learned to exercise his options well.

Sam's case well illustrates the aspect of prostate surgery that men facing it agonize over the most. If that magic pathway from the brain to the penis is obliterated, erectile functioning can suffer a devastating blow. Fortunately, surgeons who perform this operation are well aware of this hazard. Most, if not all, enter into the performance of the surgery with a deep sense of responsibility to "spare those nerves" and preserve the patient's erectile functioning.

The problem is that this is a very delicate procedure requiring skilled and experienced hands. The truth is that some surgeons are not able to do it as well as others. The very best surgeons, the ones who have had the most success at "nerve-sparing prostate surgery" have gained worldwide recognition. I know from my own experience that you don't have to have a "world class" surgeon to get great results, but it does improve your chances. Some hospitals and medical schools have developed reputations for specializing in this process, and, if you ever need them, they are easy to find.

You can appreciate the challenge facing the surgeon when you consider that he has no precise map of the microscopic nerves. Finding them is part guesswork, even for the most skilled surgeon. Fortunately, a "nerve-sensing machine" has been developed that will improve the chances of finding those important nerve bundles by every surgeon—no matter the level of his skill or experience. It's called the "CaverMap," and it was developed by the UroMed Corporation.

The CaverMap is based on nerve stimulation technology, which is already in use in delicate facial surgeries. During prostate surgery, an electronic sensor is wrapped around the penis. The surgeon can find the nerves by probing with a nerve stimulator contained in a small probe. When he stimulates a nerve, the sensor, wrapped around the penis, can sense a reaction within 15 seconds. It can sense penal reactions to the nerves being stimulated that are much too small for the surgeon to observe. Fortunately, the nerve-sensing machine detects them and it lets him know he has found a nerve by beeping. All surgeons who use the nerve-sensing machine should be able to do a better job of nerve sparing. During the surgery, when using CaverMap, the closer the surgeon's knife gets to a nerve, the faster the machine beeps in warning.

Using this technology could improve the chances of even patients like Sam of retaining near-normal erectile functioning. Although the FDA has approved CaverMap as an aid in detecting these key nerves, it is still, at the time of this writing, a very new development and not widely available for use by surgeons in hospital operating rooms. Some point out, perhaps correctly, that it is still not a replacement for a surgeon who knows what he is doing. Hopefully, the reputation of the machine as a significant aid in nerve sparing surgery will grow, and more surgeons will be willing to invest in it. Certainly, men like Sam want no expense spared in preserving their erectile functioning. So it's in the best interest of all men who are about to undergo prostate surgery to ask that their surgeon avail themselves of the CaverMap.

Marty Takes The Natural Approach To Regaining His Potency

Not long after his divorce, Marty began dating. To his dismay, after he had established a relationship with a woman and remarried, he began to realize that he was having erection problems. This prompted him to take a long hard look at himself, and, to his credit, he courageously admitted that he had some other problems as well—he was out of shape, he smoked, and he was sure he had palpitations of the heart. "I'm only thirty-seven years old," he thought "this can't be happening to me."

He was quite relieved when his doctor found nothing wrong with him, prescribed some mild sedatives and advised him to get into an exercise program. His response was to join a gym, where luckily, he happened upon the beginnings of a program both to improve his stamina and his erections. For an otherwise healthy thirty-seven year old, simply improving stamina by working-out regularly should optimize the body's production of testosterone. It will also improve his blood circulation, which leads to improved erectile functioning. It worked for Marty. But another important outcome was that he began learning, from other men in the gym, about "dynamite" things they were obtaining from the nearby health food store.

It wasn't long before he had worked out his own individualized therapy of taking botanicals and nutritional supplements. He variously included damiana, sarsaparilla, nettle root, muira puama, saw palmetto, ginseng, and lecithin. He even added zinc and L-arginine, also made sure that he got a balanced diet and he took other vitamins and mineral supplements.

One discovery that was especially helpful to him was that of heat therapy, as described in Chapter 9—"Conditioning the Penis." It was through this therapy that he convinced himself that he was fully capable of having a satisfactory erection. He also learned about the Kegel exercises and began to practice them regularly. Within six months he had all but forgotten that he had an erection problem.

There are many men like Marty with a healthy sexual interest, yet they have erection problems mainly because they are too out of shape. Furthermore, like Marty, once in shape, they may very well be able to enhance their erections by using the "tried and true" botanicals and nutritional substances that many men have relied on for years.

Steve Tapped Into The Power Of His Mind

Here we have a handsome, fit, athletic man approaching the age of thirty. He is well liked by his friends and associates, and he is polite and kind, but perhaps a bit shy. He had revealed to Dr. Bolton that he found himself making an excuse and discouraging his wife when she would, periodically, begin to intimately touch him. He had never admitted to the real reason for this either to his wife or even himself. What he and the counselor discovered together was that he did this to avoid the possibility of it going any further and leading to sexual relations. When Steve faced up to his real concern, he had to admit that he was afraid of possible erection failure.

On their honeymoon, of all times, he had had the unfortunate experience of having his erection simply "flop" at the crucial moment before insertion. No doubt this has happened to many men who simply proceed until it returns —and in most cases it does. But Steve, the polite and sensitive man that he is, was not one of those men. As you might predict, it happened in enough instances (two times more was enough for Steve) to make him truly uncomfortable at the prospect that it might happen again. It came to the point where he dreaded her advances. Steve knew he had to do *something* because he loved her very much and truly wanted to be closer to her.

Dr. Bolton realized, at once, that this was a case in which a "reprogramming" of the subconscious was no doubt the answer, and he recommended hypnosis. Unfortunately, Steve rejected this advice. He did, though, accept the advice that he see a urologist who then performed an erection test.

So, what is an erection test? Just as the name it implies, it is a test to determine whether or not a man has the ability to have an erection on his own. It is based on the hypothesis that most men have erections at various times during sleep. Indeed, clinical studies have shown that most healthy males average four to six erections a night during the deepest stages of sleep, and, most men can verify that they have had the experience of waking from sleep with an erection. Steve was given an "Inform Nighttime Erection Test."

"Inform Nighttime Erection Test" is its brand name, and usually, at least in the U.S., this test packet can be purchased in pharmacies without a prescription. In this particular erection test, a test band is worn on the penis during sleep, which is usually at night unless you work the nightshift, to determine whether or not erections are occurring. A man having ED because of psychological problems most likely will still have erections during his sleep, and the test band will indicate that he has had them. This was the case with Steve. A man whose ED is due to physical causes most likely will *not* have the normal nighttime erections.

The outcome of this was that the urologist prescribed yohimbine hydrochloride and, at the same time, asked him whether he would be interested in psychological counseling. At this point Steve found himself reconsidering Dr. Bolton's recommendation of hypnosis. He wanted to know more, so he researched the subject and finally found himself in substantial agreement with that original recommendation. But his research had also revealed to him another possibility. So instead of going back to Dr. Bolton, he decided he would try self-hypnosis.

Steve then designed his own program of self-hypnosis therapy. This program was the same as was described Chapter 10 — "The 'Mind Factor.'" He successfully reprogrammed his subconscious mind to accept his sexuality, and his erection problems went away! He credited the yohimbine hydrochloride with helping him somewhat, at first, but he soon stopped taking it. As his confidence grew, he began to notice that self-hypnosis was helping him in others areas of his life also—he was becoming more outgoing and self-assured. Later he expressed his gratitude to Dr. Bolton who had put him on the road to these successes.

Tom Found Viagra To Be His "Love Potion"

Since Viagra is on the scene, there will be many cases like Tom's. Here we have a happily married man, sixty-seven years old. He is healthy, but not particularly fit. His doctor told him to watch his cholesterol and blood pressure. There's little doubt that his libido is working just fine, but his doctor advised him that his ED stemmed from poor circulation.

Right up until April 1998, Tom had been contented with injection therapy. He had tried vacuum therapy, but reported that he found it painful to apply. But early in April 1998 he was one of the first to obtain a prescription for Viagra. Doubtful about the prospect that a pill could give him an erection, he only purchased five 100 milligram pills. But as it turned out, he reported that it was the answer to all his hopes. As he tells it, once he got the prescription filled, he and his wife, Fran, were so excited that they couldn't wait to try it, even though she was ailing with a slight cold. "It was great," he said. "We got going pretty good. I did not expect to have an erection like this again except with an injection."

Unfortunately, Fran's cold got in the way of their "hot time." She began coughing severely—so they ceased their lovemaking. Naturally Tom was frustrated, but he loved his wife dearly, so he just helped her get comfortable, then simply rolled over and fell asleep. He tells of being amazed how hard his penis was and that he even woke up the next morning with a fresh erection! Fortunately, his wife was feeling better the next morning and they made exquisite love. His climax was as good if not better than when they first got married. "Did Viagra do this?" he asked.

Results similar to Tom's are being reported everywhere. In fact, Viagra is now the most prescribed medication in the U.S.! And even though about 30% of men with ED cannot be helped by Viagra, they need not be discouraged. In all likelihood, they can use one or more of the other methods, as described in this book. There's probably a solution for almost any ED problem there is.

Chapter 12

HOW LIBIDO AFFECTS ERECTILE FUNCTIONING

"Freeing up the libido can give an added edge to any man."

What's The Most Important Element In Your Potency?

You may be surprised to learn that *desire* is the most crucial factor in erectile functioning. The stronger the desire, the stronger the erection response. This may seem contradictory, considering that men seeking to overcome ED are, in fact, men who, it must be assumed, are motivated by desire! Just the same, the beginnings of those physiological responses that result in an erection, begin with nerve signals from the brain triggered by desire. And the stronger the desire the stronger the signal.

The "mind factor," as discussed in Chapter 10, must not be overlooked in healthy men under the age of thirty-five as an element in causing erection difficulties. Unresolved guilt, irrational shame and unwarranted fear lodged in the subconscious could be a root cause of a weak desire.

Unfortunately, attitudes about sexuality exist in society, at least in the U.S., and perhaps in other countries as well, that foster mental states which give rise to guilt, fear and shame. Society at large will probably always need to instill in individuals the need for responsible control of sexual appetite, because establishing just the right amount of prudent self-discipline is, for many, a problem.

It is ironic that this need to suppress sexuality sometimes has the unwanted effect of perverting and over emphasizing it—but that is another topic. In those men whose poor erections are caused by guilt, shame or fear, one thing seems to stand out. In certain situations, usually not involving intercourse, they will have perfectly satisfactory erections. An example would be waking up from sleep with an erection.

It is believed that a great many men are affected by these "mental" interferences of erectile functioning. Very severe inhibitions require expert psychological therapy, but I would wager that most men, and women too, have inhibitions to some degree. How could it be otherwise, since everybody is subjected to a great deal of indoctrination about rights and wrongs having to do with sexuality? As a result, for many men, it is only in a very safe situation that erectile functioning operates entirely intuitively without any short-circuits in the connection between the brain and the penis.

It is not surprising then that there is a lot of lore in circulation ascribing the benefit of mind-and-mood-altering substances to improve erections. The lore and even the research in this area of speculation is, however, checkered with contradictions—one man's aphrodisiac, being another's "knockout" drops. Just the same, research has established that there actually are mood altering pharmaceuticals that have an aphrodisiac effect.

In 1987 there was a double-blind experiment, where neither the subjects nor the researchers knew which subjects were receiving the substance being tested and which were receiving something like a sugar pill i.e., a placebo. It established that bupropion hydrochloride, a drug prescribed by doctors to relieve depression, had a pronounced aphrodisiac effect on males and females. This included improved sexual functioning.[8] Other pharmaceuticals have also been reported by their users to have had an effect on their sexual functioning. The idea that such a thing is possible has been established by scientific research.

It isn't possible to advise, with certainty, that a mind-or-mood-altering substance will improve erectile functioning. Furthermore, even when it is known that a pharmaceutical has this "side effect" propensity, it is not likely to be prescribed by a physician for this purpose alone.

What About Alcoholic Beverages?

So what about the other substances with a reputation as an aphrodisiac, even though the claims about them are not established scientifically? At the head of that list are, as you might guess, some alcoholic beverages. But, I hasten to remind you as discussed in Chapter 3, which covers testosterone, alcohol's most basic effect on erectile functioning is detrimental. Despite that, it's use as a mild depressant may do enough good in removing inhibitions as to offset its negative effect in diminishing erectile functioning. Many couples have enhanced their sexual pleasure by including such things as a glass of wine in their lovemaking ritual. It is not uncommon that chilled champagne or other wine placed in the bedroom, with two wineglasses by either party, is a signal to the other for lovemaking. For the male, being careful to consume just the *small* amount needed to relax inhibitions can be an aid to improving erectile functioning.

As you might expect, experimentation with the role of alcoholic beverage in sexual pleasure has resulted in numerous ideas about what works best. One of the most widely circulated is that concerning gin. There are many who swear by gin as an aphrodisiac, given the right scenario—and the requirement that the gin not be a cheap one. The right scenario is that it be taken in a sexually stimulating situation. If used moderately by an otherwise healthy couple there is every likelihood that it will increase sexual pleasure, and thereby improve erectile functioning of the male. Taken immoderately, however, it will probably result in the couple sleeping through the entire episode, not to mention a weak erection.

How About Illegal Controlled Substances Like Marijuana?

There is considerable underground lore about the beneficial

effect of some illicit controlled substances, like marijuana, on sexual pleasure. Because of the very great possibility that a person who finds pleasure in using mind-altering substances may tend to overestimate the sexual pleasure being experienced while under their influence, the reports of sexual enhancements through using them must be accepted cautiously. Keeping in mind that reports of greater pleasure are subjective—the important question for us is whether erectile function is improved.

Marijuana is a substance often extolled for its aphrodisiac effect. Many report that it gives them a mellow, relaxed feeling. It could, in this way, set the stage for enjoying sexual pleasure by freeing the libido. There is little doubt that, like alcohol, it has the propensity to reduce inhibitions and may, in that way, improve erectile functioning. But scientific research has not established that there is anything in marijuana that could possibly be a sexual stimulant and it has no inherent potential for improving erectile functioning.

Freeing up the libido as a means to improving erectile functioning can give an added edge to men with no serious health problems and no entrenched psychological blocks. But immoderate use of alcoholic beverages will do more harm than good in improving erectile functioning. And using an illicit substance like marijuana to improve erectile functioning is going to an inadvisable extreme.

Chapter 13

SOME QUESTIONS AND ANSWERS

"Knowledge is the pathway to understanding."

Here is a list of questions many of which were brought to my attention by interested readers of the pre-publication edition of the manuscript for this book, along with the answers which may be helpful to you.

Question 1: *For seven years I have been successfully using the vacuum device my doctor prescribed after the removal of my prostate. My surgeon tells me that he was successful in sparing the nerves that were attached to my prostate and connected to my penis, and that therefore I should be able to attain an erection somewhat like I did before the surgery. But my erections have been satisfactory only with the help of the vacuum device. Do you think Viagra will help me?*

Answer: You are probably an excellent candidate for Viagra. Since you were able to achieve a partial, though weak erection, the nerve connection between your brain and your penis is not entirely severed. With Viagra the weak response in your penis should, under sexual arousal, be strengthened. All other things being equal, you may be amazed and entirely pleased with the results.

Question 2: *I am glad that Viagra is here, and it really works for me. I had been functioning on partial erections most* *of the time for years. My wife has never complained. But since I started taking Viagra, I can tell by her new reactions that she has noticed, and now appreciates the difference that Viagra has made. Our lovemaking is better. I am forty-eight years old, but I feel like I am nineteen again. Can you tell me how taking a pill can make that happen?*

Answer: It's complicated, but the medical scientists explain it something like this: Under sexual stimulation, meaning you have to be "turned-on", something very important happens in the penis. You probably only experience it as, perhaps, an urge—for some, an overpowering urge—but that "important something" is that *nitric oxide* is being released in those two spongy chambers of the penis—the corpus cavernosum. Here is where it gets interesting. The nitric oxide causes production of something called cyclic GMP (cyclic guanosine monophosphate) to go into high gear.

Don't be alarmed. These strange sounding chemicals occur naturally in the body. Having sufficient amounts of cyclic GMP is absolutely crucial to having an erection. The problem is that another naturally occurring body chemical (phosphodiensterase) can break down or degrade cyclic GMP as fast or nearly as fast as the nitric oxide can cause them to appear. In a man with no ED problem, nitric oxide will win out. If he has no blood flow problems, the man will have sufficient cyclic GMP, and he will have an erection.

What Viagra does is stop the breakdown of cyclic GMP by inhibiting that sinister chemical that can prevent an erection. In effect, Viagra stops phosphodiensterase from breaking down, or destroying, cyclic GMP. Cyclic GMP can then do its wonderful job of relaxing the smooth muscle arterial tissue so that the penis can fill up with blood and thereby squeeze shut or pinch off the little "hoses" that let the blood out again. As long as the cyclic GMPs are doing their job, a man will maintain an erection. But once sexual excitement has passed, thus shutting off the production of nitric oxide, the production of cyclic GMPs stops. Then the remaining amounts of cyclic GMPs are soon degraded, and the penis returns to a flaccid state.

The marvelous thing about Viagra is that it works only on stopping that chemical, phosphodiensterase, in the penis that breaks down cyclic GMP—in effect it's a "magic bullet." But your body has to produce the nitric oxide on its own. For Viagra to do you any good, you must be sexually stimulated. If you fail to get "turned-on," at least enough to produce some cyclic GMPs, Viagra will have nothing to do and will, in a few hours, be harmlessly excreted from the body. If you do get "turned-on," Viagra will enable whatever amounts of cyclic GMPs you are able to produce to have maximum effectiveness. In this way, it enables men to overcome weak or no erections.

Question 3: *So if you don't get sexually aroused, Viagra will not help you achieve an erection. What can you do about that? Is there something you can take to make you feel sexually stimulated?*

Answer: There are actually pharmaceuticals that seem to do just that, but as a side effect. If this is a libido matter, you may want to discuss it with your physician—especially on the outside chance that your testosterone level is too low. But review some of the natural herbs and nutrients discussed in this book, such as damiana, sarsaparilla, nettle root, muira puama, saw palmetto, ginseng and lecithin; they may be all you need.

Question 4: *Recently I began a relationship outside of my marriage of five years. I have a very satisfactory sexual relationship with my wife though I do not feel happily married. I never have any problem with erection failure except in my new relationship. I had complete failure. The funny thing is I felt really turned-on but I didn't get an erection hard enough to finish what we had started. And it has happened more than once. I am afraid to get in this situation again. What's going on here? Is there any therapy you would recommend for me?*

Answer: Since you didn't say anything about your physical condition, I will assume that you were not tired or otherwise stressed. This is not an uncommon occurrence when the subconscious mind has accepted a suggestion of "guilt," in that you are performing or about to perform an action that your conscience is telling you that you *should not*. You may have been violating principles that you had consciously or more likely, subconsciously accepted.

If you do this you are living a lie and are incongruent with yourself. This can lead to a lot of psychological problems, not to mention legal and moral issues.

In any case, you need to examine your belief system. See Chapter 10—"The 'Mind Factor.'" Search the depths of your psyche and ask yourself if this act is in harmony with your belief system, and are you violating your own principles if you violate the pledge of fidelity that you no doubt made to your wife? Your subconscious mind may be disrupting that all important signal from the brain to the penis if you are about to act in disharmony with your deep belief system. You may discover that your own principles, and the friendship and companionship that you and your wife share, commands the higher priority in your life.

Question 5: *I am one of those guys you wrote about "who just wants to get it harder." How can I get a vacuum device or injections for occasional use? Will a doctor prescribe for mild erection problems? And what about cost? I don't have health insurance.*

Answer: It is very likely that a doctor will try to help you, but only after a physical examination, to assure that you are a suitable candidate for these therapies. This will usually include actually self-administering these therapies in the presence of a doctor. Finding the right doctor is getting easier every day, particularly if you live in a larger metropolitan area. In some of these areas they advertise almost daily in newspapers typically announcing help with or cures for impotence. Also, manufacturers and pharmaceutical companies maintain toll free telephone lines through which you can be referred to physicians who prescribe their products. Finally, the yellow pages in your telephone book are a source. Look for Physician Referral Services.

Since the introduction of Viagra, many doctors who formerly did not treat ED have begun accepting and treating patients seeking to obtain Viagra. Understandably, since Viagra has been proclaimed "to work," it is a legitimate response to a patient with a complaint of erection problems to immediately issue him a prescription for Viagra. However, at the same time that you are seeking help from a physician, you would be well advised to take inventory of your lifestyle and physical condition. Remember that exercise is the key to blood flow and good testosterone levels.

As for cost, these treatments are relatively expensive—particularly, if you do not have health insurance coverage. I would predict, if you are interested in Viagra, that in time the price will drop, especially if other pharmaceutical companies are successful in producing rival oral medications that work.

In the meantime, in my opinion, the least costly and most reliable treatment for occasional erection failure is vacuum therapy. However, in this case too, without health insurance, the initial outlay will be somewhat considerable. It will entail a consultation with a doctor and a visit to the pharmacy to present your prescription and purchase the appliance. The cost, not including the doctor's fee and depending on which manufacturer your doctor selects, and whether you select a battery or manually-operated pump, the cost could range from $175 to over $300 (U.S. dollars).

The other pharmaceuticals (injection and urethral suppositories), without health insurance coverage, will start at about $80 a prescription. That may sound like a better price, but keep in mind that you receive a very limited number of applications and the medication has a limited shelf life. The vacuum appliance is reliable and most manufacturers will replace the pump mechanism if it fails, without cost.

Question 6: *I am in that less than 30 age group and I don't have any problem getting an erection except when I really need one. Do you understand? What would you advise?*

Answer: I take it you mean that you have erection failure just as you are about to have sexual intercourse. This is a situation that obviously calls for counseling from a qualified sex therapist. However, a physician could prescribe injection therapy or vacuum therapy and this would assure an erection. This might be just the nudge you need to become comfortable enough to overcome your obvious anxieties. As a guiding principle, I would say that it is always better to use your body's natural resources if at all possible, and that "nudge" might be enough to allow you to do just that. But if the "nudge" does not lead to a breakthrough, then counseling will certainly be called for.

Another approach that could be helpful is reciting positive affirmations as discussed in Chapter 10—"The Mind Factor." Affirmations could lead to freeing you of inhibitions that are limiting your life. You might find it enlightening to engage in a thorough psychological self-examination. This could be challenging to accomplish on your own. A behavioral therapist could very possibly be of great benefit to you in such an undertaking. There is at least one thing you could do on your own that might very well give you clues to the answers in solving the psychological interferences with your erectile functioning: examine your dreams and the thoughts you fall asleep with. Or better still, see a dream counselor. Yes, there actually are psychological counselors who specialize in "dream counseling". They look to find answers to explain behaviors and disease through examining the content of an individual's dreams.

Question 7: *What are the side effects of using injections?*

Answer: Virtually none, for most of the prescription pharmaceuticals, such as prostin, now being used in injection therapy. Rare instances of priapism, erections that won't go down or continuous or long-lasting erections, have been documented. In this case, your doctor will have to be consulted as soon as possible to apply an antidote. Another concern for users of injection therapy is the possibility of scarring, usually because of incorrect or careless use of the needle. Scarring is, of course, not a side effect of the medication itself, and it has not been a big problem. Luckily, for men who absolutely wish to avoid the sting of a needle, however slight it may be, there is an alternative method of applying this type of pharmaceutical (see Chapter 6—Urethral Suppositories). But, you may be a candidate for that miracle of modern pharmacology, Viagra.

Question 8: *Some men seem more at ease with women than* *others. Does this have any effect on erectile functioning?*

Answer: An erection is more easily aroused in a man who is relaxed and without anxieties. It is a rare man who can function sexually while experiencing nervousness at the same time. Men who are at ease in the company of women normally have the advantage of not placing psychological roadblocks in the path of that signal from the brain that initiates the process of getting an erection. This applies to women who likewise are ill at ease in the company of men; they too can experience roadblocks to arousal even though they may want to experience it. This is not an unusual problem for newlyweds. In either case, it is advisable to seek professional counseling, particularly if the problem is severe. Shyness, or being extremely ill at ease in the company of the opposite sex, is not something you need to live with.

Question 9: *What effect will using these therapies have on my ability to have an erection naturally on my own in the future? If I use these injections continually over a long period of time will they wipe out any chance of my ever regaining the ability to have an erection naturally in the future? Also, is injection therapy addictive?*

Answer: No, you won't be forever hooked. At least not in the sense of being physically addicted. Some men have overcome periods of temporary ED through these methods and have not experienced the need to continue. In some cases, they overcame psychological barriers and in other cases, they improved their erectile functioning through health and exercise practices such as I have described.

While it is a comforting feeling for a man with ED problems to know that these therapies have the potential to make him fully functioning should he need them, having the ability to function immediately upon arousal certainly has its charms. Otherwise healthy men would however, do well to improve their blood flow and stamina by exercises as described earlier. They should also reject practices detrimental to their health such as smoking and excess drinking of alcoholic beverages.

Question 10: *If I use injection and get an erection that won't go down, what do I do?*

Answer: Call your doctor, you'll need an antidote. Before you receive this pharmaceutical, your doctor is expected to have informed you of this possibility. As mentioned earlier, this condition is called priapism. Though rare, it can occur even without using injections. The good news is, the incidence of priapism caused by using injection therapy is low.

Question 11: *What about scarring of the penis through use of injection therapy? Is it a serious problem?*

Answer: So far it has not been what could be characterized as a serious problem. Very few men have experienced scarring. In an even smaller number, the scarring has resulted in a curvature of the penis. Carefully observing the proper technique for injection will, in nearly all cases, result in the ability to have a very satisfying sexual experience with no undesirable side effect, such as scarring. However, if you are unfortunate enough to incur serious scarring, your doctor may have to take you off injection therapy.

Question 12: *How do I find a doctor who treats ED?*

Answer: Many doctors specializing in urology also treat ED. To find one, first ask your family doctor or personal physician for a referral. However, if you don't have a family doctor or personal physician, there's a very good chance you can find the doctor you need by simply looking in the telephone directory yellow pages. If you live in a large metropolitan area you will probably be able to easily locate a clinic that specializes in treating ED problems. Look for advertisements in leading newspapers offering treatment for "impotence." In addition, the companies which manufacture pharmaceutical and other treatments for ED maintain toll free numbers to refer prospective customers to doctors who prescribe their products. Also, see "Appendix B" of this book. The information sources listed there may be helpful.

Question 13: *I am age 70, and because of prostate surgery, I will probably never attain an erection on my own again. As a result, I am relying on a vacuum pump. Are there any potential negative side effects from long-time use of this device?*

Answer: None are expected. Of course, caution is always

advisable to not exceed the vacuum pressure necessary to achieve a good erection. Otherwise, your main consideration needs to be to maintain healthy circulation to assure good blood flow to the penis. You can do that by maintaining an aerobic exercise program and otherwise following good health practices. The low impact walking program previously discussed (see Chapter 2) should be adequate as a maintenance program. This, of course, assumes you are not involved in behavior detrimental to your health such as smoking or excessive use of alcohol.

Question 14: *I have heard of a treatment that only involves massaging a cream onto the penis. I am told that the treatment originated in China. Could this possibly work?*

Answer: I, too, have heard of it or perhaps a similar product. The product you refer to is probably one called "Tenfold." The manufacturer claims that it is a traditional Chinese medical ointment applied with local massage. Supposedly, it is manufactured in the People's Republic of China and reportedly it is very expensive to purchase in the U.S. As you might expect, the manufacturer claims a very high success rate in treating ED.

Could this possibly work? It is certainly conceivable for some cases of ED. For instance, testosterone deficiency is being treated by medical doctors in the U.S. by applying a patch to the scrotum. But I know of no research data on this treatment, and I remain skeptical. As I understand it, the treatment requires applying the massage technique over a period of time in order to achieve a cure. If it really does work, you will surely eventually, and maybe sooner, hear of it from some conventional sources.

Question 15: *Will any of the treatments discussed in this book have any influence on reproductive capacity?*

Answer: None expected. Children have been conceived by men using virtually all the therapies discussed. On the other hand, no man is going to become fertile as a result of using these therapies either. Hopefully he will become a better-prepared participant in the sexual aspect of his relationship, but not more fertile. Likewise, he will probably not improve his refractory time—he'll still need his normal recovery interval between orgasms. These therapies can improve the erectile functioning of the men who need to use them, and for those not wishing to impregnate, birth control must be practiced.

Question 16: *You have listed a lot of herbs and nutrients. Would you please break it down to a simpler list of things that work best to improve erections? In other words, what items would I need to put on my shopping list for a visit to a health food store?*

Answer: Are you asking, "Isn't there something...for guys who just want to get it harder"? If so, I'll bring to your attention a few natural substances that have recently received increased attention and are in higher demand. Although the men who use them experience what seems to them rather immediate and verifiable results, it'll be up to you, through self-monitoring, to decide if they work for you, too.

The list includes *Muira Puama*—it reportedly has a dynamic effect on sexual desire and performance with no side effects. If claims for this botanical are to be believed, it increases the body's natural production of testosterone.

Withania somnifera—an Indian herb reputed to be excellent for treatment of sexual deficiency in men —is now found in many health food stores and is worth a try.

DHEA (delhydroepiandosterone) has recently become a very popular supplement and with good reason. Its effect on the body is similar to that of testosterone and it has a following as an aphrodisiac. It promotes an overall feeling of well being and it may increase libido. It is, after all, a hormone, and like testosterone, is produced naturally in the body. Although DHEA is in vogue and being consumed in ever increasing amounts by both men and women, men should use caution in taking it. Over use could, in time, slow the body's natural production of it. There is also concern, by some medical doctors, about possible adverse effects on the prostate gland that could possibly result from prolonged use, which is the same kind of concern they have about taking testosterone.

Also include on your shopping list these venerable old standbys: *saw palmetto* and *sarsaparilla,* possibly formulated as *smilax officinalis. Yohimbe* is another substance that has proven reliable for many men, but I would recommend that you get a prescription from your doctor and obtain pharmaceutical grade yohimbe. Some men have experienced undesirable side effects from some "health food store" formulations. In all cases, verify that you are purchasing a quality-manufactured item of the highest purity available. Other than that, make sure you are maintaining proper nutrition, exercise and rest. And, of course, check with your doctor.

Question 17: *I have learned of a device that enables a man to have intercourse with a partial erection. It is available in drug stores without a prescription. It is called "Rejoyn" and seems to be available everywhere. It is even advertised on the radio. Is this possible?*

Answer: Such a device may actually be a good idea for some men, if it is acceptable, of course, to their wives. I believe it falls into the category of devices sometimes referred to as erection collars or erection supporters.

It is sometimes possible, as perhaps more than a few men know, to "stuff" a semi-erect penis into a vagina. The purpose of this device, as I see it, is to make the process easier, and thereby, facilitate intercourse. But these devices do not cause or produce an erection. Instead they enable intercourse by supporting a partially erect penis.

Rejoyn is an example of a support sleeve or collar. It's described as "a soft and comfortable support sleeve that fits around the flaccid penis, giving it the rigidity needed for sexual intercourse."

Question 18: *Could Viagra be affected by other drugs?*

Answer: Yes! For instance, if you're on nitroglycerine, you could pass out. Tagamet, on the other hand, could increase the effects of Viagra! In general, make sure it's compatible with any medications you're currently taking by checking with your doctor. It might also, as an extra precaution, pay you to ask your pharmacist—especially if you're under the care of more than one physician, both or all of whom prescribe medications for you.

Question 19: *Can Viagra help a veinous leak?*

Answer: It might help overcome the adverse effect of a mild leak, since Viagra's effect is to cause a greater flow of blood into the penis than flows out.

Question 20: *Are there any side effects to taking Viagra?*

Answer: Some men have experienced side effects which are discussed in Chapter One in some detail. But for the otherwise healthy male, the transitory side effects to beware of include that the color green and blue may become somewhat indistinguishable and that some men may experience headaches and/or nausea. These are among the mild and tolerable side effects. Doctors are coming to realize that the likelihood of experiencing them is increased with taking the higher dosage pills, as you might expect. These side effects, as I stated, are some of the mild and tolerable ones. But if you have any kind of heart or circulation (or even respiratory) problem, the possible side effects are much more serious. In such cases, you must consult with your doctor.

Question 21: *When should I take Viagra?*

Answer: It's best taken about an hour before you plan to have sex. If taken on an empty stomach, it can be effective in as little as fifteen minutes. If taken after a heavy meal, it may not have an effect for two or three hours. So, if you and your lady return from a great restaurant dinner, don't expect to "pop" a Viagra and have it immediately kick in. To be effective, it has to be absorbed into the bloodstream, and it may have to compete with the fine meal you have just eaten.

Question 22: *Is Viagra for everyone?*

Answer: No, first of all, not all men are fit candidates for Viagra. Secondly, it works for 70% of men and not for the rest. Be sure to thoroughly discuss your health history and all the medications you're on, if any, with your physician before you use it. In fact, it's obtainable only by prescription. Refer back to Chapter 8 for more details about whether you should use Viagra or not.

Question 23: *Can Viagra help with premature ejaculation?*

Answer: It may offer a form of treatment for this condition;

but there's no conclusive evidence. Remember, Viagra is a serious prescription drug that is to be used to treat a serious medical condition. Accepting some pills from a neighbor or friend, "just to try it," would not be wise! See your doctor, who may recommend treatment, if needed—Viagra or something else.

Question 24: *Are there situations where taking Viagra is dangerous?*

Answer: Yes! Anyone on heart medications, like those containing nitrates, should not take Viagra. The interaction can cause one to pass out and possibly lead to death. Anyone with high or low blood pressure could be at risk, as well. Also, anyone who has not performed sex for some time, and is not used to that level of physical activity, could be at risk by taxing the heart and causing a heart attack. In any case, consult with your physician before making any changes in your routine.

ED is a serious medical condition that needs the attention of a doctor, and it could be an indication of a serious health problem. Be sure to inform your doctor, as well as your pharmacist, about all the medications you are taking. The easiest way to do that is to bring your prescription bottles when you visit your doctor and pharmacist.

And remember, Viagra is a prescription drug. Don't obtain it or use it unless a proper diagnosis has been done and a prescription has been written specifically for you.

If your doctor says Viagra is not for you, perhaps he'll prescribe or recommend one of the other ED solutions presented in this book.

"The Second Coming Of 'The Pill'"

Years ago, the birth-control "pill" gave women, who were fertile, more control over their childbearing potential, and it still does. Now Viagra and several other solutions give men more control over their ability to perform. Finally, a concern of men since the beginning of time, has been addressed and dealt with. Use these solutions wisely and with discretion. And now, a final word...

A Final Word

"For nearly all men, ED is a problem they can do something about."

The Answer To Larry's Question is YES!

The answer to Larry's question, and one that many men with ED have, is certainly a resounding YES. Given that you don't have a serious illness that prevents it, you can overcome ED and achieve "potency-on-demand"!

It is my hope that this book will be helpful to men seeking to restore and enhance their erectile functioning as well as those who just want to maintain it as many years as possible. Fortunately, we live in a time when a subject like this can be discussed without embarrassment, and we're also fortunate that there's an abundance of information available from a lot of different sources. And even as these pages are being assembled, new products and information are making dramatic appearances.

In 1998 many believed that we had seen the ultimate triumph over ED with the introduction of "the impotence pill" (Viagra—sildenafil citrate from Pfizer). It effectively enables 70% of all men to overcome erectile dysfunction simply by swallowing a pill. For nearly all men, ED is a problem they can now do something about. As we go into the 21st century, ED will be successfully treatable by many variations of the therapies discussed in this book–oral, transurethral and injection agents, as well as creams, ointments and more. Some are even speculating that gene therapy will be used to treat ED.

Stay Active And Involved As You Age

As I have stated elsewhere, the triumphs over ED made since the early 1990's are but one aspect of continual push to extend the limits of human potential. It is an area of human striving whose time seems to have arrived. John Glenn, that great American hero, personifies this spirit. At age 77, Glenn, who was the first man to orbit the earth in 1962, trained for a return to space in 1998. His many accomplishments as a military hero,

Lieutenant Colonel in the United States Marine Corp, and as a United States Senator certainly would justify a vacation for the rest of his life. But at age 77, John Glenn "is promoting a more personal realization: that human beings continue to be worthy, interesting and capable, as they grow old."[1]

In the magazine, *Men's Health* of June 1998, on pages 132-137, it tells of an 83 year-old who regularly participates in barefoot-water-skiing, skydiving, snowboarding and runs marathons. The article also features a 91 year-old who still runs marathons, and a 65 year-old bodybuilder with such a youthful looking body that I would challenge anyone to guess his age. The article is titled "Live to 90 And Die Having Sex." The recommendations of Joe Kita, the author of that article, for staying healthy are entirely consistent with those I make in these pages for maintaining and improving erectile function.

Assess Your Health and Fitness And Take Appropriate Action

Perhaps this is the best time to say, particularly to those men who have had the recognition of their empowerment renewed by Viagra, that this is the time to take an assessment of your general overall health and total being. What areas of your health, you relationships, your lifestyle, and your mind can or "should" you strive to improve. In particular, if you are going to reap maximum benefit from taking Viagra, improve your heart and circulatory system. Get the blood flowing where it is needed. And remember the Kegel exercises!

With the introduction of Viagra, more men than ever are overcoming ED. The remaining question is whether such powerful pharmaceuticals are the appropriate answer for all men in general. Are they suitable for the many individuals for whom only modest improvement in heart and circulatory health is all that is needed? Many men could gain that extra edge by finding and exploring the herbs and botanicals available in health food stores. Still others could benefit by adjusting their attitudes, mentality and outlook, and perhaps a lot more by simply adopting a healthier lifestyle!

It can be argued that the clamor for Viagra by so many younger men, as a means of erectile enhancement, should be taken as a clear call to persuade younger men to embrace more health conscious practices. This could enable them to naturally retain much, if not all, of their youthful sexual vitality well into their older years, and perhaps for their entire lifetime.

As we discussed, blood flow and exercise are the keys to natural healthy erectile functioning. A man in his 20's or 30's, or even his 40's, needs to strive first for the optimum physical condition that promotes optimum blood flow and testosterone levels. In younger men, the potential for vibrant erectile function, triggered only by sexual arousal, is a great gift to be appreciated and not just taken for granted. It presents them with a spontaneity that is the hallmark of youth. They need not squander it, but rather, diligently preserve and maintain it!

For the rest of us, say 50 or older, and those of us who've had whatever degree of ED, we may restore and continue or maintain our maximum sexual performance as, perhaps, an extra gift to ourselves and our mates. That gift perhaps, results from our own wisdom or the natural inclination of mankind to ever expand the limits of what we can achieve.

There's Hope For All Of Us...

Another fine example of a couple overcoming obstacles to preserve the treasured intimate aspect of their marriage is that of Christopher Reeve and his wife Dana. Chris is known as the star of the *Superman* movies. In May 1996 he was thrown form his horse and sustained a serious spinal cord injury, leaving him paralyzed from the neck down. He's been on a ventilator and in a wheelchair ever since, but he didn't let that stop him. As a result, in fact, he's written the book *Still Me*. He's also a movie director, does some acting and is a strong advocate for and helps raise funds for medical research to help people with paralysis.

Chris and Dana were featured in a segment of ABC television's *20/20*, which aired on May 1, 1998. Among other things, Dana talked about the honest, open, loving communication they have with each other. At the conclusion of the interview, Barbara Walters asked Dana whether she and Chris are still able to have relations. I don't recall the exact words of Dana's response, but she smiled and said something to the effect that they can get very creative.

In the final analysis, we are examining a very complex human response. Consider a letter and response that appeared in the syndicated "Ann Landers" advice column that came out on the heels of the triumphant announcements of the "impotence pill." In her advice column of April 16, 1998, as it appeared in the *Los Angeles Times,* and presumably all across North America, a wife writes into respond to a man who had written a letter earlier, which was printed in a previous column. He had said that no woman would ever want him because of his "sexual impotence." In an encouraging reply to him, she told of the successful sexual relationship she has with her husband, who is paralyzed from the chest down, the result of a tragic automobile accident. She states, "We were a very sexually active and loving couple before the accident, *and we still are today* (emphasis, mine)."

The wife goes on to say "His doctors recommended implants and other devices, but we opted for our own creativity. This was not an easy transition for me. I thought our marriage was over if we couldn't maintain our sex life. My therapist, however, made me realize that there was more to our marriage than sex. I chose to stand by my husband and adjust to this tremendous change. It turned out to be very wise and rewarding decision...If you have a caring and loving relationship, you can handle whatever comes your way."

This woman and her husband triumphed over devastating physical impairment that included ED—and have done so, remarkably, *without pharmaceutical or therapeutic aids!* This story and that of Christopher Reeve and his wife Dana are the ultimate affirmations that ED *can* be overcome.

Ann Landers' reply says it all: "...the most important sex organ is the brain. ...A lively imagination, creativity and a sense of humor can be enormously helpful."

Always remember that the human potential for triumph

includes three greater tools than anything discussed so far—*courage, love and resolve.*

And for you, dear reader, armed with the information presented in these pages, exercising your imagination, creativity and sound judgment, you too, can demonstrate "potency-on-demand."

It has been my endeavor to bring together the information that men need to find the solution to their ED problems. Be prepared to make your own investigation and evaluation of what solution(s) work the best for you, including the new products that are coming out and will continue to come out. So, if this book points a man, or the woman who loves him, to a source or solution that proves helpful or useful, I am gratified that I have done at least some of what I intended. If I reach just one couple who achieve a healthier, happier, stronger marriage, I will have succeeded.

May you reap the rewards of overcoming ED—restoring your sexual functioning or improving it and maintaining it for a life time. May it enrich and enliven your relationship, positively affecting the quality of your life and that of the woman who loves you, contributing to your overall well being.

"We must have courage to bet on our own ideas, on the calculated risk, and to act. Everyday living requires courage if life is to be effective and bring happiness."

–Maxwell Maltz

Appendix A: Performing The Kegel Exercises

Kegel exercises, practiced regularly, are the single most important exercises you can perform to improve erectile functioning. You'll also gain other important benefits like control of your orgasms, more pleasurable, and possibly longer-lasting orgasms and better bladder control (ladies can achieve this last benefit too). However, they cannot substitute for aerobic and weight training exercises—especially low impact aerobics and the "walking program" described in Chapter 4. An optimal degree of stamina and endurance is essential.

Remember, your "PC" is the muscle you clench to stop the flow of urine or to push out the last few drops. For your convenience, here, again, are the Kegel exercises instructions from Chapter 3. Use the chart in this section for your initial 6 to 8 week conditioning period and for maintenance afterward. Feel free to reproduce this chart on a copier.

How To Do This Exercise

Begin by squeezing your "PC." Hold this for about 10 seconds and then relax. SQUEEZE AND RELAX. You should start with your "PC" relaxed and then gradually squeeze as hard as you can, holding it for 10 seconds. Do this for at least 15 minutes or more, if you still feel comfortable. As you gain control, you will find many opportunities to do the exercise, like when you're sitting.

Another variation is "quick snaps" of the "PC". These are done in the same way but quickly. SQUEEZE AND LET GO TO A RHYTHMICAL BEAT, like music or even your own heartbeat.

Still another variation is the SQUEEZE AND HOLD technique—squeeze your "PC" as hard as you can and hold it for up to 30 seconds. Squeeze even harder just before you let go. Then relax for 15 seconds. Continue these repetitions for a total of 15 minutes.

Be sure to do the Kegel exercise in a methodical manner. Use the chart to plot and organize your exercises, and plan on doing them twice a day for 6 to 8 weeks. Thereafter, you may want to continue a less intense maintenance program. The potential benefits from a dedicated Kegel exercise program can be a tremendous aid in your achieving and maintaining peak sexual performance.

Week__ These Kegel Exercises can improve your erections and increase your pleasure!				
DAY	**Number of quick snaps**	**Number of squeeze & holds**	**Time of AM Exercise**	**Time of PM Exercise (at least 6 hours after AM)**
Monday				
Tuesday				
Wednesday				
Thursday				
Friday				
Saturday				
Sunday				

Appendix B: Information Sources

Not Surprisingly, Pfizer's internet address is one of the busiest addresses:

Pfizer's Internet Address
http://www.viagra.com/

For More Information on Viagra For Men:
Men's Consultation Clinic at Johns Hopkins Hospital
410-955-6707

For More Information on Viagra Research for Women:
Women's Sexual Health Clinic at Boston University
617-638-8555

Injection Therapy:

Caverject®

Call 1-800-867-7042 (in the U.S.) for free information and the name of physicians in your area who have prescribed this pharmaceutical. In Canada, call 1-888-678-2820.

Web Site for information about injection therapy:
http://www.pharminfo.com/pubs/msb/caverject.html
Also: http://impotent.com/whatis.html

Urethral Suppository:

Muse® (Alprostadil)
1-888-367-MUSE
http://www.vivus.com

Vivus, Inc.
545 Middlefield Road
Menlo Park, CA 94025

Vacuum Therapy:

Obson Medical Systems (ErecAid System)
In the U.S. call 1-800-344-9688 or 1-800-438-8592.
In Canada call 1-800-356-4676.
http://www.osbon.com

Stokes Medical Arts Pharmacy, Inc. has a web page that includes a feature on vacuum therapy (Post-T-Vac): http://www.stokesrx.com/vacuum.htm.

VET-CO, INC. (Vacuum Erection Technology System) on the internet at:
http://www.jax-ads.com/impotence/vetco.htm
(or call 1-800-827-8382)

A List Of Sources Of Related Information

Dr. Andrew Weil's web site: http://www.drweil.com (An excellent source of health information of all kinds with a decided emphasis on alternative medicine and natural products).

Dr. Isadore Rosenfield–his latest book is: *Dr. Rosenfeld's Guide to Alternative Medicine: What Works, What Doesn't and What's Right for You.*
His articles appear in *Parade* a magazine that is distributed in newspapers nationwide. (1998 Parade Publications, 711 Third Avenue, New York, NY 10017-Chairman and Publisher Cario Vittorini).
"A Doctor's Guide to Herbs" appears in May 31, 1998, pp12-13.

For future articles: "Do you have questions about health or medicine? Send them to Dr. Rosenfeld, and he will try to answer them in future articles. Personal replies are not possible. Write to: Ask Dr. Rosenfeld, P.O. Box 5099, New York, NY 10017.

Christopher Reeve Foundation
P.O. Box 277, F.D.R. Station, New York, NY 10150-0277; 888-711-HOPE
Christopher Reeve Foundation internet address:
http://ww.apacure.com

Sexual Function Health Council
c/o American Foundation for Urological Disease
1128 North Charles Street
Baltimore, MD 21201-5559
800-242-2383

Impotence Institute of America (IIA) & Impotents Anonymous
8201 Corporate Drive, Suite 320
Landover, MD 207115
800-669-1603

National Kidney and Urologic Diseases
Information Clearing House
3 Information Way
Bethesda, MD 20892-3580
301-654-4415

American Diabetes Association
National Center
1660 Duke Street Alexandria, VA 22314
800-232-3472

Sexual Function Health Council
American Foundation for Urologic Disease, Inc.
1128 North Charles Street
Baltimore, MD 21201
Toll-Free Telephone Number 800-242-2383

Sexual Function Health Council, a council of the A.F.U.D.
For information on the treatment of ED, call: 800-727-8984, or access this Web Site Address: http://www.access.digex.net/~afud/takes2.html

Glossary

Aerobic Exercise the activities that exercise the heart and the cardiovascular system.

Alternative Medicine the treatments generally not considered to be traditional, based on the idea that natural herbs and remedies pose much less risk to the overall well-being of the body than the potent chemicals resulting from medical research.

Aphrodisiac food, drug, potent or other agent that arouses physical desire for sex.

Erectile Dysfunction (ED) problems getting and sustaining an erection.

Erection upon sexual arousal, the brain sends a signal to increase blood flow into the penis while at the same time decreasing the flow of blood out, thus causing the penis to become engorged with blood and rigid.

Hypnosis an artificially induced mental state like sleep, characterized by heightened susceptibility to suggestion.

Impotence the inability to achieve and sustain an erection suitable for intercourse; erectile dysfunction (ED).

Injection Therapy a pharmaceutical, usually a form of prostaglandin, that ultimately increases blood flow is injected directly into the penis, to cause an almost-immediate erection.

Libido your sexual desire, which affects your erection response.

"Mind Factor" the affect that the ideas held in the subconscious mind, even from childhood, have on the ability to have an erection. Reprogramming is required.

Natural Remedies composed of, derived from or by products of foods that provide a multitude of health benefits.

Nitrates a salt of nitric acid contained in heart medications, as in nitroglycerine.

Nitric Oxide a complex chain of proteins that, when present, cause the relaxation of the blood vessels in the penis that produce an erection.

Nutritional Supplements something that can be eaten or swallowed that act, in treating ED, as an aphrodisiac.

Positive Affirmations replacing your harmful beliefs with helpful ones, by reprogramming the subconscious, in the case of ED, that support healthy erectile functioning.

PC (Pubococcygeus muscle) a muscle that contracts during ejaculation. When Kegel exercises are done, as discussed in Chapter 3, it improves blood flow to the penis causing harder erections, and more powerful and pleasurable orgasms. It can also be used to control premature ejaculations.

Penile Implant a surgically implanted device into the penis to enable a man to perform sexual intercourse.

Physical Fitness improving the physical condition of your body; In treating ED, especially those parts of the body that have a direct influence on your ability to achieve an erection.

Potency a man's capability of performing sexual intercourse.

Prostate Gland a walnut-sized gland that wraps around the urethra at the base of a man's bladder which controls the release of urine and makes up part of the semen.

Sildenafil Citrate the active ingredient of the pill trademarked Viagra by Pfizer, an oral therapy for ED.

Subconscious that part of your mind that is generally out of reach, but in the case of ED, with which you will learn to communicate in order to restore or enhance your erections.

Testosterone the male hormone which is essential in the proper functioning of the libido. Without it, the signal from the brain that produces an erection doesn't reach the penis.

"Training Effect" the point during exercise when the heart is required to exert nearly maximum effort to keep the body, particulary the large muscles, supplied with oxygenated blood.

Urethral Suppositories like injection therapy, only without the needle and syringe. Uses a substance similar to prostaglandin called alprostadil in the form of very small suppositories which are inserted into the urethral opening at the head of the penis through which urine and semen flow. When absorbed, it initiates an erection.

Vacuum Therapy using a specially-designed vacuum device, by means of a clear plastic cylinder placed around the penis, to which is attached a small pump. A vacuum is created around the penis, causing blood to flow into the penis for a firm erection.

Viagra Pfizer's trademark for sildenafil citrate, an oral therapy for ED.

Citations

1 *Los Angeles Times*, page E2.

2 *Newsweek*, November 17, 1997. Front Page and 62-8.

3 *Los Angeles Times*, October 22, 1997, page A3, "Herb May Slow Alzheimer's," Thomas H. Maugh II (Times Medical Writer).

4 "New State TV Ads Link Smoking to Impotence in Men," by Dan Morain, *Los Angeles Times*, June 2, 1998 pp A3 & A16.

5 "Experience with buccal phentolamine mesylate for impotence," Zorgniotti, A.W., Dept. of Urology, New York University School of Medicine, NY 10016, *International Journal of Impotence Research*, 1994 Mar, 6:1, 37-41.

6 *Los Angeles Times* Editorials, "Sex and the Insurance Man," p B10, April 27, 1998.

7 "A Pill for Impotence?," *Newsweek* (Lifestyle), pp. 62-68, Nov. 17, 1997.

8 Cited by Crooks & Baur, *Our Sexuality*, 4th edition, p 189, The Benjamin/Cummings Publishing Company, Inc. 1990.

9 Ann Landers, "Love, Sex Survive Accident," *Los Angeles Times*, p E4, April 16, 1998.

"...Love is
the fury of the storm
the calm in the rainbow
Love is the source of passion..."

Susan Polis Schutz

Bibliography/Suggested Reading

The American Medical Association Family Medical Guide, Random House, Inc. New York, NY, 1982.

Balch, James F., M.D. and Phyllis A. Balch, C.N.C., *Prescription Nutritional Healing, Second Edition,* Avery Publishing Group, Garden City Park, NY, 1997.

Caprio, Frank S., M.D. and Joseph R. Berger, *Helping Yourself With Self-Hypnosis,* Prentice-Hall, 1974.

Clapp, Larry *Prostate Health in 90 Days Without Drugs or Surgery,* Hay House, Inc., Santa Monica, CA, 1997.

Covey, Stephen R., *The Seven Habits of Highly Effective People,* Simon and Schuster, New York, NY, 1989.

Fuhrman, John, *Reject Me-I Love It!,* Success Publishers, Pennsylvania, 1997.

Helmstetter, Shad Ph.D., *What To Say When You Talk To Yourself,* Simon and Schuster, New York, NY, 1982.

Maltz, Maxwell M.D., F.I.C.S., *Psycho-Cybernetics,* Simon and Schuster, New York, NY, 1971.

Mandino, Og, *The Greatest Secret in the World,* Bantam Books, New York, NY 1975.

McGrane, William Jr., CPAE, *Brighten Your Day With Self-Esteem,* Success Publishers, Pennsylvania, 1996.

Peale, Dr. Norman Vincent, *The Power of Positive Thinking,* Ballantine Books, New York, NY, 1982.

Reeve, Christopher, *Still Me*, Random House, New York, NY, 1998.

Rifenbary, Jay, *No Excuse-I'm Doing It!*, Success Publishers, Pennsylvania, 1997.

Rosenfeld, Isadore, *Dr. Rosenfeld's Guide to Alternative Medicine: What Works, What Doesn't and What's Right for You*, Random House, New York, NY, 1997.

Shames, Richard M.D., and Chuck Sterin, M.S., Ph.D., *Meditation and Self-Hypnosis*, Rodale Press, Emaus, PA, 1978.

Index